Musabekov Aidos

PHARMACEUTICAL BIOTECHNOLOGY

Musabekov Aidos

PHARMACEUTICAL BIOTECHNOLOGY

Textbook

ScienciaScripts

Imprint

Cover image: www.ingimage.com

This book is a translation from the original published under ISBN 978-620-7-45070-1.

Publisher:
Sciencia Scripts
is a trademark of
Dodo Books Indian Ocean Ltd. and OmniScriptum S.R.L publishing group

120 High Road, East Finchley, London, N2 9ED, United Kingdom
Str. Armeneasca 28/1, office 1, Chisinau MD-2012, Republic of Moldova, Europe
Managing Directors: Ieva Konstantinova, Victoria Ursu
info@omniscriptum.com

Printed at: see last page
ISBN: 978-620-8-52775-4

Contents

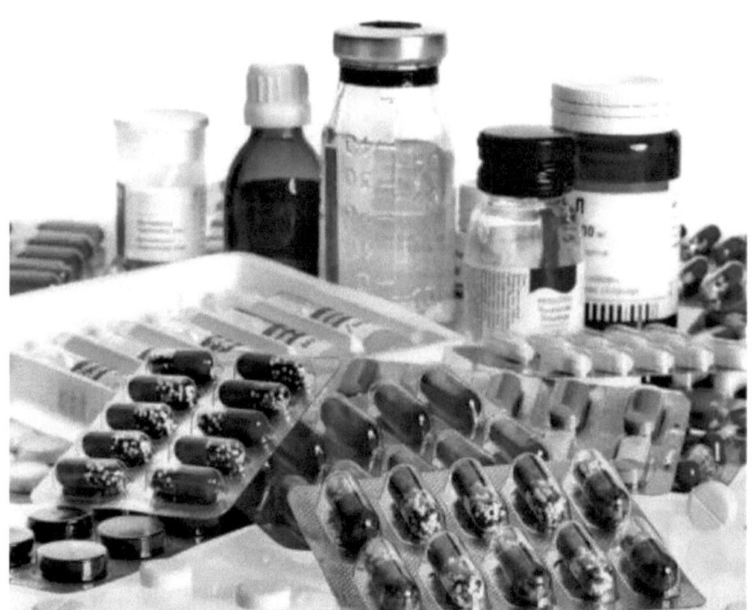

The features of biotechnological processes of obtaining the main types of drug substances used in medicine, veterinary medicine and other areas of national economy, as well as methods of their isolation, purification and fractionation are considered. The procedure of creating regulatory and technical documentation for products of pharmaceutical biotechnology is analysed, their characteristics are given.

Textbook for students of higher education institutions studying in the speciality 6B05102-Biotechnology.

PREDICTION

Pharmaceutical biotechnology is the most important section of modern biology. At the end of the 20th century, it has become one of the priority areas in international science and economics.

Pharmaceutical biotechnology is an essential

The development of research activities in the field of pharmaceutical biotechnology took place in the 1980s, when new methodological and methodological developments began to be applied in science and practice, and there was a real opportunity to make the most of it from the economic point of view. According to forecasts, biotechnological products are expected to account for a quarter of the world's production in the 21st century.

In the 1980s, the research base in Belarus was significantly expanded and its data began to be actively used in production. During this period, the first national programme on biotechnology was developed and implemented in the country, interdepartmental biotechnology centres were established, qualified biotechnologists were trained, and biotechnology laboratories were set up in research institutions and universities.

However, the attention to the problems of biotechnology as a whole in the country weakened, and the volume of financing in this direction decreased. This led to a decrease in the level of development of biotechnology as a science in Belarus, especially in such a field as genetic engineering.

As for more modern biotechnological processes related to the production of medicinal substances, they are based on methods of recombinant deoxyribonucleic acids (DNA), as well as on the use of mobilised enzymes of cells or cell organelles. Modern pharmaceutical biotechnology is the science of genetically engineered and cellular methods and technologies for the creation and use of genetically transformed biological objects in order to intensify production or obtain new types of products for various purposes.

The manual deals with the main objects, general principles of biotechnological processes, areas of application of biotechnology products, methods of industrial production of purified biologically active substances of biotechnological origin in industrial biotechnological production.

Vizdanin accumulates the basic principles of biotechnological production of the main groups of pharmaceutical substances, purification of biologically active compounds such as proteins, amino acids, vitamins, organic acids, lipids, nucleotides and nucleic acids, sugars and polysaccharides, antibiotics, alkaloids, various fermentation products, hormonal preparations, enzymes, food additives, antiserums, as well as the main directions of their practical application.

Currently, the microbiological industry uses thousands of strains of various microorganisms. In most cases, they have been improved by induced mutagenesis and subsequent selection. This makes it possible to synthesise various substances on a large scale.

Some proteins and secondary metabolites can be obtained only by culturing eukaryotic cells. Plant cells serve as a source of a number of compounds: atropine, nicotine, alkaloids, saponins and others. Animal and human cells also produce biologically active compounds: for example, pituitary cells produce lipotropin, a stimulant of fat breakdown, and somatotro- pine hormone, which regulates growth.

The created grafted cultures of animal cells producing monoclonal antibodies are widely used for diagnostics of diseases. In biochemistry, microbiology, and cytology, methods of immobilisation of both enzymes and whole cells of microorganisms, plants, and animals are of undoubted interest. Such biotechnological methods as cell and embryo culture, *in vitro* ovogenesis, artificial fertilisation are widely used in veterinary medicine. All this shows that biotechnology is a source not only of new food products and medical preparations, but also of energy and new chemicals, as well as organisms with specified properties.

Chapter 1

DEVELOPMENT OF THE PHARMACEUTICAL INDUSTRY IN THE WORLD IN THE XX C.

The pharmaceutical industry is an industry that develops, manufactures and markets licensed drugs and medicines. It is characterised by a variety of legislative and government regulation with respect to patenting, testing, safety and efficacy of the medicines produced.

The emergence of pharmacies dates back to the Middle Ages. The first of them opened in Baghdad in 754, then they spread rapidly in medieval Arab countries and Europe. By the 19th century, many pharmacies in Europe and North America had become large pharmaceutical companies. Most of today's pharmaceutical companies were formed-in,,to,,too late 19th_(early 20th)-.

In the 1920s **and** 1930s**, insulin and penicillin were discovered and became** drugs produced on a large scale. During this period, the pharmaceutical industry was most developed in Sweden, Germany, Italy, the United Kingdom, the United States, Belgium and the Netherlands. This was also the time when legislation was developed to regulate the testing and approval of medicines, requiring the use of appropriate brands. It became possible to legally separate prescription and over-the-counter medicines.

In the 1950s, systematic scientific research was conducted, medical knowledge and the technological process of production were improved. During this period, a large number of new drugs were developed and massively distributed, including **cortisone** and various cardiac agents. In the 1960s, tranquillisers and psychotropic drugs such as **chlorpromazine, chaloperidol, diazepam** appeared and found extremely wide use.

At the same time, attempts were made to strengthen state regulation and limit the financial ties between pharmaceutical companies and prescribing physicians. In particular, the American Food and Drug Administration (FDA) was created. It turned out, for example, that the uncontrolled use of one of the new tranquillisers **(thalidomide)** by pregnant women led to the birth of children with various deformities.

In 1964, the International Medical Association adopted the Declaration of Helsinki, which set standards for clinical trials and required pharmaceutical companies to prove the efficacy of new drugs under clinical conditions before launching them on the market.

The 1970s was a period of anticancer drug development. Until the 1970s, the pharmaceutical industry remained a relatively narrow sector of the economy. It is from this time that a period of rapid development begins. Most of the countries adopt stringent patent laws. Since 1978, India has become a centre for manufacturing pharmaceutical products without patent protection.

By the mid-1980s, small biotechnology companies in their struggle for survival began to actively merge with large pharmaceutical corporations or sell their shares to them. There was a process of rapid liquidation of pharmaceutical companies themselves, the largest of which took a dominant position not only on the national pharmaceutical markets, but also on the global one. In the 1980s, safety and environmental legislation was tightened, and new medicines aimed at combating HIV and heart disease became a hallmark of the decade.

The 1990s saw a wave of mergers and acquisitions in the pharmaceutical market and a dramatic increase in contracts with research organisations for clinical trials and basic research and development. The marketing process has changed significantly. The Internet has enabled direct purchases of medicines by consumers and of raw materials by drug manufacturers, which has changed the nature of business. In the USA, with the adoption of new legislation liberalising risk requirements in 1997, advertising of medicines on radio and television became widespread. A generation of antidepressants, including the most popular fluoxetine, has emerged. So-called alternative medicine and nutritional supplements began to develop rapidly, which increased competition in the pharmaceutical industry. At the same time, pharmaceutical companies were increasingly criticised by the public for trying to "impose" new diseases on the population and, consequently, drugs to "fight" them.

To better understand the functioning of the pharmaceutical industry, it is necessary to refer to the concept of "drug discovery". Drug discovery refers to the process by which potential drugs are designed. In the past, most drugs emerged from the separation (isolation) of active ingredients from traditional medicines or from accidental discoveries. Modern biotechnology concentrates on investigating the metabolic processes that occur during a particular disease or pathogenic condition and utilises molecular biology and biochemistry. The early stages of new drug development have traditionally been carried out by universities and research opra- low!(ions).

Drug development begins once the basic ingredients of a drug have been identified and aims to find a suitable formulation and dose, as well as the safety of the drug. Research in these areas usually includes in *vitro, in vivo* and clinical trials. As the major stages of new drug development require significant investment, the process is carried out by large pharmaceutical companies.

Multinational corporations are often vertically integrated, operating simultaneously in different segments, from research and development of new drugs to manufacturing, quality control, marketing, sales and distribution. Smaller companies often focus on specific aspects of pharmaceuticals, such as the development of individual drug components or drug formulations.

The process of drug discovery and development is extremely capital-intensive. Of all the drug ingredients under investigation, only a small fraction pass the government approval stage and are authorised for use. Globally, on average, only 25 new medicines receive marketing authorisation each year.

Such authorisation can only be obtained after huge investments in preclinical development and clinical trials, as well as in ongoing monitoring of the safety of medicines. Some medicines cannot undergo this procedure and do not provide a return on investment. If we take into account the costs of development of such drugs, the cost of creating one successful drug (including the costs of its marketing and distribution) may reach USD 2 billion. Such estimates are rather conditional, as they do not take into account the costs of regulation, state subsidies, tax incentives and federal research grants. This is why the real costs of bringing new medicines to the market are considerably higher.

In the United States, new pharmaceutical products are reviewed for safety and efficacy by the Food and Drug Administration. This process involves providing a large amount of preclinical information about a new drug, as well as three stages of clinical trials. First, the toxicity of the drug is determined using groups of healthy volunteers. At the second stage, pharmacokinetics, routes of administration and dosage are worked out. In the third, the longest, the effectiveness of the drug on specific patients is determined. Often a fourth, post-marketing stage is added to ensure the safety of the medicine and record side effects.

The United States has special rules for creators of medicines for rare diseases, which affect about 200,000 people in the country. Since the cost of research and development of such drugs is extremely high and financially unprofitable, the United States incentivises pharmaceutical companies through tax breaks, subsidies and exclusive market access for such drugs for a certain period of time (usually seven years), regardless of whether such drugs are protected by patents™. not.

In 2006, the total value of the global pharmaceutical market was estimated at $640bn, of which the US accounted for almost 50%. The pharmaceutical industry remains one of the most profitable sectors, with a sales margin of

17%. The world's best-selling drug is *Pfizer*'s cholera-sterilising tablets, which had annual sales of 13 billion shares in 2008, more than *doubling* to 13 billion. 13 billion share, more than double the sales of Plavix, a cardiovascular drug *from Bristol-Myers Squibb*, and the anti-asthmatic drug Advera from *GlaxoSmithKline.*

Table 1 shows data on the largest global pharmaceutical and biotechnology companies. All of them make up the so-called *Big Pharma* group, к which includes companies with sales of over $3 billion and R&D expenditures of over $500 million.

Table 1

Largest global pharmaceutical companies (2008)

Company	Country	Sales volume, billion USD	R&D volume, billion USD	Number of employees, thousand people
Novartis	Switzerland	53,3	7,1	138
Pfizer	U.S.A.	48,4	7,6	122,2
Bayer	Germany	44,2	1,8	106,2
GlaxoSmithKline	UK	42,8	6,4	106
Johnson & Johnson	U.S.A.	37,0	5,3	102,7
Sanofi-Aventis	France	35,6	5,5	100,7
Hoffmann-La Roche	Switzerland	33,5	5,3	100,3
AstraZeneca	UK	26,5	3,9	50,0
Merck & Co.	U.S.A.	22,6	3,9	74,3
Abbott Laboratories	U.S.A.	22,5	2,3	66,8
Wyeth	U.S.A.	20,3	3,1	66,7
Bristol-Myers Squibb	U.S.A.	17,9	3,1	60,0
Eli Lilly and Company	U.S.A.	15,7	3,1	50,0
Amgen	U.S.A.	14,3	3,4	48,0
Boehringer Ingelheim	Germany	13,3	2,0	43,0
Shering-Plough	U.S.A.	10,6	2,2	43,0
Baxter International	U.S.A.	10,4	0,6	38,4
T jkeda ... . " Pharmaceutical Co.	Japan	10,3	1,6	15,0
Genentech	U.S.A.	9,2	1,8	33,5
Procter & Gamble	U.S.A.	8,9	n.s.v.	29,3
Total		497,5	70,8	1342

Any pharmaceutical company can apply to the relevant government authorities and obtain a patent for a drug or its manufacturing process with exclusive rights, which is usually valid for 20 years. However, this procedure involves very rigorous testing, inspections, which takes 10-15 years, after which the company is authorised to market and sell its drug. Patent protection enables the patent owner to recover research and development costs through high profits from the sale of branded drugs. After patent expiry, competing companies usually develop so-called *generic* medicines, which are less expensive to create and less expensive to approve, allowing them to be sold at lower prices. Often, the company that owns the branded drug starts to produce the generics before the patent expires in order to capture the market.

The market in this area is characterised by a high intensity of mergers, acquisitions and co-operations that exploit the complementary capabilities of companies. Small pharmaceutical firms may possess new drugs but lack sufficient sales."" marketing.capacity. On the contrary, large corporations often have idle capacity in this area. Consequently, both are looking for joint opportunities to increase capitalisation through the synergies of cooperation.

Pharmaceutical companies typically spend significant amounts of money on advertising, marketing and lobbying for their products. In the US, these companies spend about $20bn annually to market their medicines. Advertising is done through medical journals and media channels. In some countries, including the US, such advertising is allowed to go directly to the public. Pharmaceutical companies usually employ specialised staff to promote medicines to doctors in medical institutions. In some countries, most notably the United States, companies lobby politicians to promote their interests. The marketing of prescription drugs in the USA is regulated by a special federal law passed in 1987.

Health professionals and medical practitioners are among the most important players in the pharmaceutical market, as they write the prescriptions that determine which medicines patients will use. Influencing physicians is a key element of the prescription drug sales process.

A medium-sized pharmaceutical company usually has about 1,000 sales representatives; in large corporations their number reaches tens of thousands of people. In the USA the number of sales agents is about 100 thousand. Only for four years (1999-2003) the number of agents has doubled. Pharmaceutical companies annually send about 5 billion dollars to pay for the work of their representatives with medical institutions.

The pharmaceutical industry is increasingly at the centre of controversy and public criticism. Some of the reasons for this include: companies exerting pressure on doctors and medical institutions through sales representatives, including the provision of 'marketing gifts' and biased information; aggressive advertising in journals and at conferences; funding of independent medical institutions and wellness campaigns; lobbying of doctors and politicians (especially in the USA); sponsorship of medical schools and treatment courses; funding of ongoing educational seminars with the aim of improving the quality of life of doctors and medical practitioners; and, in some cases, the use of the pharmaceutical industry as a vehicle for the promotion of healthy lifestyles.

According to independent estimates, the pharmaceutical industry spends more money on lobbying than any other sector of the US economy. Between 1998 and 2006, companies spent about $1 billion on lobbying. Pharmaceutical companies are often accused of "disease selling" (overuse of medicines) to expand the market for medicines, as well as of providing misleading information to health professionals who end up prescribing expensive medicines to patients when other medicines may be cheaper and more effective.

Figures 1, 2 show the positive and negative aspects of public perceptions of the pharmaceutical industry in China, Japan, Mexico, South Africa and the United States.

Modern science has great potential for the development of new drugs. Today, people in developed countries live on average 30 years longer than they did a century ago. Reduced mortality and significant improvements in the quality of life are largely to the credit of the pharmaceutical industry.

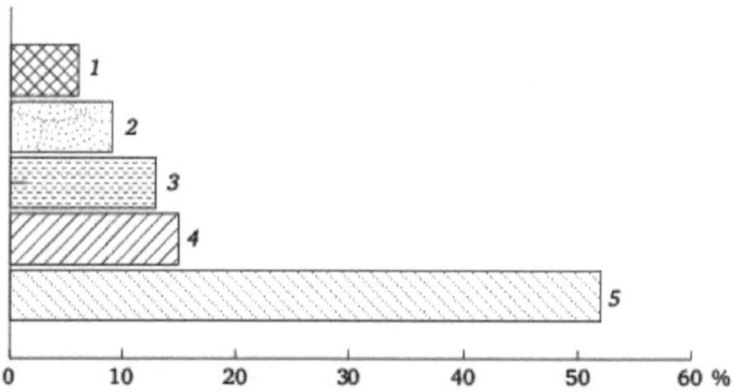

Figure 1. Most important positive characteristics of the pharmaceutical industry, per cent surveyed:

1 - co-operation with the medical community (6 %); *2* - improved life expectancy (9 %); *3* - quality and effective medicines (13 %); *4* - efficient production process and distribution (15 %); *5* - high level of innovation, production of new medicines (52 %).

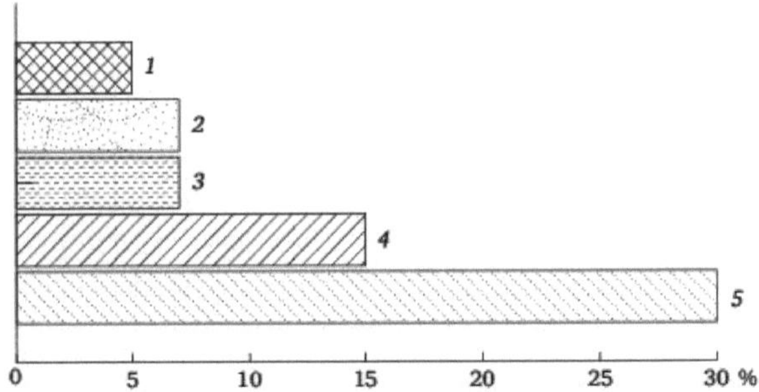

Figure *2*. Most important negative characteristics of the pharmaceutical industry, percentage of respondents:
1 - aggressive marketing (5%); *2* - lack of transparency regarding risks and clinical trials (7%); [3] lack of social responsibility and ethics (7%); *4* - focus on profitability and more expensive drugs at the expense of cheaper drugs (15%);
5 - high prices and inaccessibility of medicines (30%)

A key contribution of the pharmaceutical industry to medical progress is the use of fundamental research to develop innovative medicines. Since then, scientific and technological advances in the pharmaceutical industry (primarily through research into the human genome and individual cell composition) have made it possible to successfully combat many complex diseases and identify their causes.

All new medicines that appear on the market are the result of a long, expensive and risky research and development process conducted by pharmaceutical companies. The cost of bringing new medicines to market has grown exponentially over the past decades. As can be seen in Fig. 3, over thirty years, the total cost of bringing new drugs to the pharmaceutical market in developed countries has increased almost tenfold. The high risk of failure, the cost of lengthy clinical trials, and the rising costs of obtaining government approvals are the main factors behind their rapid growth.

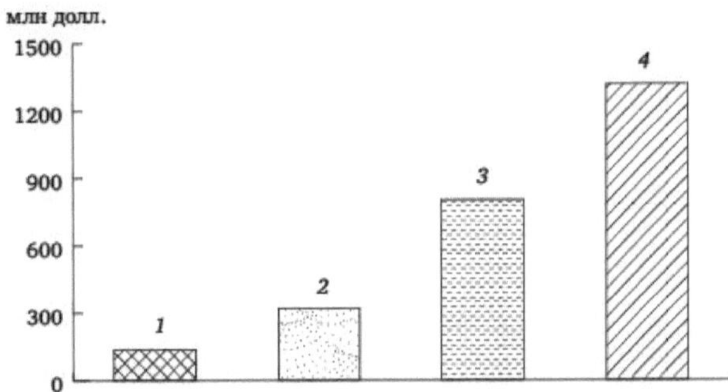

Figure 3. Cost of bringing new drugs to market:

1 - 1975; *2* - 1987; *3* - 2001; *4* - 2006.

Many promising substances, after reaching a sufficiently advanced stage of clinical trials, are then rejected. The percentage of drugs that reach the marketable stage remains very low (according to various estimates, 1-2 per 10,000). Financing such research and development requires significant amounts of cash, which a pharmaceutical company is able to generate only if it brings new drugs to several national markets as soon as possible. In other words, such companies must inevitably be global.

The research and development process in the pharmaceutical industry consists of several stages. The first stage is the obtaining of a patent and the start of preclinical research. Its duration is usually 4 years. Then comes the clinical trial stage, which includes three phases and lasts about 7 years. Finally, the last phase (or fourth phase), which lasts 3 years, is the marketing authorisation, pricing and other administrative procedures. The whole process takes an average of 13-15 years.

Of the total R&D expenditure, which in pharmaceutical companies amounts to 18-20 % of sales, approximately 27 % is spent on preclinical research. Almost 54% is spent on clinical trials, 5% is spent on obtaining various approvals from state authorities and 14% on additional trials required after obtaining approval documentation (Table 2).

For many years, Europe held the leading position in the development of new drugs. However, since the mid-1990s, the role of the leader in the pharmaceutical market, as well as in biotechnology, has been taken over by the USA (Fig. 4).

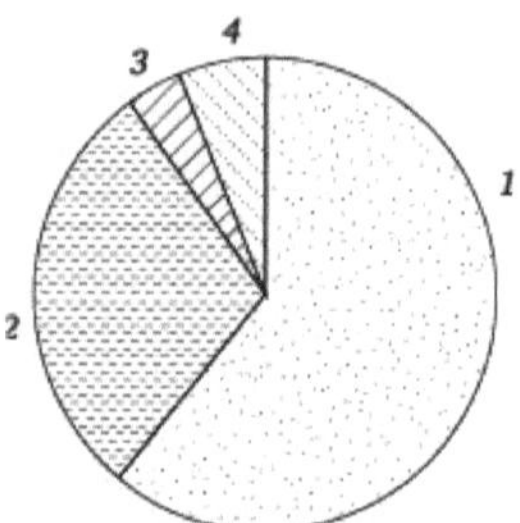

Figure 4. Sales of medicines launched in 2005-2009, by region, %:
1 - USA; *2* - Europe; *3* - Japan; *4* - other countries

As we can see, the USA accounts for 60% of new medicines, Europe 29%, and Japan 4%. This is due to the fact that in recent years in the USA a much larger share of GDP has been allocated to research and development than in

European countries. The USA has spent 2.76 per cent and 1.90 per cent of GDP on research and development respectively. Among the European countries, only Finland (3.75 per cent) and Sweden (3.73 per cent) have a higher relative level of expenditure.

The leading position of the USA is determined primarily by the more developed biotechnology, which is the core of modern pharmaceuticals. The application-

Structure of R&D investment by stage of the pharmaceutical process

Table 2

Stage	Number of patients	Duration	Table of Contents	Share of total investment in R&D, %
Pre-clinical testing		4 years	Synthesis of new substances, biological screening, pharmacological testing	27
Clinical trials including:		It's only 7 years		54
phase 1	20-100 healthy volunteers	Up to 1 year	Toxicity testing, safety testing. Selection of optimal dose	8
phase 2	Several hundred patients	1-2 years.	Evaluating effectiveness and identifying side effects	13
phase 3	A few hundred to a few thousand patients	Ages 2-4	Confirmation of efficacy and side effects with long-term use	33
phase 4 (additional tests)	Usually a few thousand patients	Varies	Identification of new users, comparison with other medicines, determination of the clinical effect and long-term safety of the medicine on a large sample of patients and compliance with the conditions of the authorisation documentation	14

he use of knowledge of the human genome in clinical practice and in the development of new medicines makes it possible to anticipate a patient's reaction to a particular drug and to create new medical "personalised" medicines in accordance with the peculiarities of the human genetic structure. The introduction of such medicines not only reduces the incidence of disease, but also changes the very model of medical care for patients, shifting the focus to preventive treatment.

Comparative data on the level of biochemistry development in different countries and regions are given in Table 3.

Table 3

Biochemical sector development indicators (2008)

Indicator	Total	U.S.A.	Europe	Asia
Sales volume, billion USD/%	59,6/100	45,0/75,5	11,2/18,8	3,4/5,7
Ob? TM.TM(vestments) ... in R&D, $bn/%	21,1/100	17,2/81,5	3,5/16,6	0,4/1,9
Number of employed, thousand people/%	192,7/100	128,2/66,5	49,0/25,4	15,5/8,9
Number of companies, pcs/%	704/100	371/52,7	178/25,3	155/22

Table 3 shows that, in terms of biotechnology development, the U.S. for the The United States is the world's leader and has an overwhelming advantage over its competitors. However, in terms of research efficiency and expenditure, the US lags behind both Western Europe and Asia: while these regions have a smaller share of R&D expenditure, their share of sales is significantly higher (for example, the ratio for Asia is 1.9% and 5.7%, respectively).

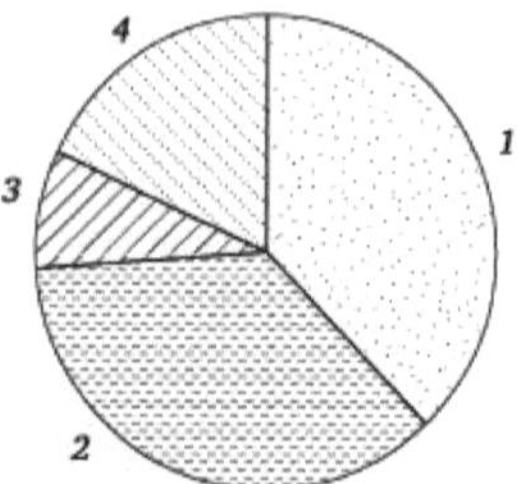

Figure 5. Regional structure of the world pharmaceutical . "...PRoduction: - *1* - USA; *2* - European Union;
3 - Japan; *4* - other countries

The USA is the world's leading producer of pharmaceuticals (38%), ahead of Europe (36%) and Japan (8%). These three regions account for 82% of global pharmaceutical production (Fig. 5).

According to international statistics, the pharmaceutical industry is the most knowledge-intensive and innovative sector of the world economy: it has the highest net output per employee and the highest ratio of R&D expenditures to sales. This sector occupies the main positions in the US industry structure (tab. 4).

Table 4

Sectoral structure of R&D in industry (2008), %

Industry	U.S.A.	European Union	Japan
Pharmaceuticals	25	17	8
Computer manufacturing	24	13	12
Software and services	15	3	2
Automotive	9	25	27
Electronic	2	5	13
Chemistry	3	6	7
Others	22	31	31

As we can see, the three most knowledge-intensive industries in the US account for 64% of all research and development, while in Europe - 33% and in Japan - 22%. This position ensures the US global technological leadership.
Developed countries with a strong pharmaceutical industry do not provide direct government support to the industry, but actively stimulate the development of new areas of research: biotechnology, genetic engineering and others. In the USA, for example, state support of the industry is provided mainly at the regional level through stimulating the development of natural sciences, biotechnologies and innovations. For example, in the mid-2000s, the state of Texas allocated almost $1 billion to expand existing and create new research laboratories in the field of biotechnology. In New York State, the New York State Science, Technology and Innovation Support Centres (*NYSTAR*) have funded more than 20 educational institutions serving the needs of the pharmaceutical industry.
Each state uses one or another of the traditional methods of tax support: tax credits, tax holidays, tax reductions, etc. Special discounts on medicines and import regulations are also used. Public-private investment funds are used to support R&D. For example, the Bioscience Institute in North Carolina is a venture capital fund that uses public funds and private investors to finance scientific and research opra- tions. The California State Pension Fund provides $500 million annually for a biotechnology R&D programme. Data banks have been established in each state to provide information, support and advisory services to R&D organisations and small and medium-sized businesses.

Developing countries, for their part, are fully supporting pharmaceuticals in all areas: access to finance, investment in production, human resources, infrastructure, and investment in research and development.

Active state support of the pharmaceutical industry has allowed a number of developing countries, primarily India and China, not only to protect domestic markets, but also to successfully enter foreign markets. The share of own manufacturers in the pharmaceutical market of China is currently 70 %, and that of India - 80 %. These countries have contributed to the growth of their role in the world market. Thus, the share of Indian and Chinese companies in the global pharmaceutical market increased from 1999 to 2007 from 5.5% to 9.4%.

In India, support measures for the pharmaceutical industry have evolved from supporting manufacturing to capture the domestic market to encouraging R&D to enhance the country's competitiveness. The development of government support measures for the pharmaceutical industry in India can be divided into several phases. Between 1948 and 1960, the pharmaceutical market was largely dependent on imports. The state support was only in the organisation of state-owned companies. In the 1960s, the state began to support the production of pharmaceutical substances necessary for the resource supply of the industry, which allowed the subsequent development of its own pharmaceutical industry.

In the 1970s, after the signing of the *Patent Act*, which abolished patenting of products, foreign companies lost their monopoly on the drugs they had developed and began to withdraw from the Indian market. The domestic market continued to develop at the expense of its own manufacturers, who copied well-known pharmaceuticals.

In the 1980s and 1990s, the successful development of India's manufacturers became the basis for a "drug policy" aimed at Indian pharmaceutical growth and international expansion. In the 1980s and 1990s, the successful development of Indian manufacturing became the basis for a 'drug policy' aimed at in- teasing growth and international expansion of Indian pharmaceuticals through aggressive export promotion. In 2000, World Trade Organisation (WTO) membership commitments forced India to reintroduce pharmaceutical patenting. But at the same time, a 'pharmaceutical policy' was developed to support R&D as a critical lever to maintain India's competitive position in the global market. Research and development support funds were actively established.

State support measures affected the whole range of human resources, production incentives, R&D and the formation of the necessary

infrastructure.
The system of educational grants and support of educational programmes for training researchers and managers for the pharmaceutical industry have been developed in the field of human resources.
Measures to stimulate production included reducing the list of drugs for which a price ceiling is set, reducing profit tax, financing export operations and preferential loans, searching for foreign partners and promoting products on foreign markets. In the area of R&D stimulation, support was provided for joint research programmes of the industry, educational and research institutions. State grants totalled up to 70% of the cost of each project. Preferential loans were provided to companies involved in R&D, international research organisations were involved in working with Indian partners, import duties on R&D equipment were reduced, and small and medium-sized innovative companies were supported through a special body, the *Small Business Innovation Research Initiative.*
Pharmaceutical industry infrastructure includes the establishment and maintenance of a database of ongoing research, information on all export support programmes through the *Export Promotion Cell in the Pharmaceutical Division*, biotechnology parks and incubators.
Active government support has enabled Indian pharmaceutical companies to protect the domestic market and significantly increase exports. Today, India's own manufacturers account for 80% of the pharmaceuticals market, up from 30% in 1970. The volume of exports increased from USD 0.7bn in 1996 to USD 3bn in 2006. Today, the Indian pharmaceutical industry consists of 20,000 companies and employs 500,000 people. In 1996-2008, its growth rate (17%) was twice as high as that of the world market (9%). The share of Indian manufacturers in the global drug market increased to 2.2 per cent by 2015 compared to 1.6 per cent in 2008.
In China, state stimulation of the pharmaceutical industry has evolved to support the competitiveness of Chinese manufacturers while liberalising the domestic market and entering global markets. Until 1980, the pharmaceutical industry was completely under state control. In the 1980s and 1990s, in connection with the transition to a market-based economy, laws were adopted to regulate the production of medicines and their quality, patenting of products was introduced, and a special *State DragAdmrnvtratton* Agency (later transformed into the *State Food and Drag Administration*) was established to centrally regulate the production, trade, registration, certification and quality control of medicines. In 2000, in connection with China's accession to the WTO, intellectual property legislation was tightened

and the distribution system was liberalised. A unified quality system was introduced. In 2006, a programme for innovative development of pharmaceuticals was adopted with unprecedented investments and the objectives of the industry's transition from copying Western drugs to creating its own innovative drugs and developing R&D.

In the area of human resources, emphasis was placed on the development and support of educational programmes to train researchers and managers for the pharmaceutical industry, including grants and support for the development of "pharmaceutical schools", the number of which increased from 5 in 1952 to 240 by 2006. In the area of production incentives, grants and favourable loans for production expansion and working capital financing were provided. A stricter quality control system has been introduced and taxes have been reduced to promote Chinese products for export. The government also promotes the registration of Chinese medicines in other countries.

State support for R&D is provided through the promotion of technology imports, preferential loans to private companies conducting research, and tax incentives for foreign pharmaceutical companies establishing joint ventures with Chinese companies and opening their own research centres in China.

The state support of the pharmaceutical industry allowed Chinese companies to increase their sales volumes at a rate outstripping the growth of the world market (annual growth was 20%) and to almost double their share in the global pharmaceuticals market in ten years (from 5% in 1998 to 8.5% in 2008).

The Russian pharmaceutical market is one of the most dynamic in the world. According to experts' forecasts, it will continue to grow rapidly in the coming years and may become one of the three largest markets in Europe. In 2003-2008, the volume of the Russian pharmaceutical market grew from $4.3 to $17 billion (on average by 32% per year), and its share in the global market of pharmaceuticals increased to 2.2%. The high growth potential is determined by the fact that Russia lags three times behind the world average in per capita consumption of medicines (USD 53 and USD 160, respectively). However, Russia does not have its own large pharmaceutical companies. The existing ones are being forced out of the market by foreign manufacturers. Thus, the share of Russian manufacturers in the domestic pharmaceutical market decreased from 28% in 2003 to 20% in 2008, while the share of imports increased from 72% to 80%. Russian companies practically do not export medicinal препараТ№ In 2007, exports of pharmaceutical products from the country totalled 0.04% of global sales. By comparison, the export volume of India (whose market is smaller than

Russia's) was 0.6% of global sales, i.e. 15 times higher than Russian exports. Russian companies are mainly engaged in the production of low-value-added products. In the structure of consumption of innovative drugs on the Russian pharmaceutical market, domestic production accounts for only 3%, while 97% is imported. And only in the consumption of non-branded generics the share of domestic products reaches 52%.

The low competitiveness of Russian companies compared to foreign ones (both in the foreign and domestic markets) is primarily determined by insignificant production volumes. For example, the largest Russian pharmaceutical company Pharmstandard has sales volumes of less than USD 0.6bn, while the global Swiss corporation Novartis has sales of USD 53bn. R&D budgets are not comparable at all: the five largest Russian companies spend a combined $15-20 million on R&D, while Novartis spends $7 billion.

Chapter 2

PHARMACEUTICAL DEVELOPMENT OF INDUSTRY IN THE REPUBLIC OF BELARUS

The pharmaceutical industry is one of the sectors of the Belarusian economy that has been developing most dynamically and at a high rate recently. This is due to many factors. Let us mention some of them.

Firstly, this sector contributes to the national security of the Republic of Belarus in the field of health care, medicine and medical and technical support.

Secondly, the pharmaceutical industry is one of the knowledge-intensive, innovative areas of the economy, the development of which is supported by the White Russian.government and the state as a whole.

Thirdly, the development of the entire national healthcare system depends on the effectiveness, availability, and proper quality of medical products.

Fourthly, in the world pharmaceutical industry is a highly profitable sector of the economy, net profit in which, reaches 18% of the total income (the average for the economy - 5%), and sales volumes and growth rates of the pharmaceutical market are constantly increasing. There are objective reasons for this situation. On the one hand, these are global trends of population ageing, increasing income levels and quality of life, and on the other hand, the well-known side effects of economic development based on scientific and technological progress (increasing impact of technogenic factors, deterioration of the ecological environment, etc.), which inevitably lead to an increase in morbidity.

During the years of sovereignty of our country, the pharmaceutical industry has undergone dramatic changes. In the 1990s, there were two pharmaceutical companies operating in the territory of the Republic of Belarus. In the 1990s, there were two pharmaceutical companies operating in the territory of the Republic of Belarus and now there are more than 20. All of them are united in the concern "Belbiopharm". Among them are RUE "Belmedpreparaty", OJSC "Borisov Medical Preparations Plant", RUE "NesvizhskiiZMP", JV "Minskintercaps", JVOO "Farmland", UE "Dialek", RUE "Exon", RUE "Izotron Plant", RUE "Grodno Medical Preparations Plant", LLC "Pharmtechnology", RUE "ENZIM" and others. According to expert estimates, the volume of the Belarusian pharmaceutical market in 2009 in estimated selling prices totalled about USD 600 million. The share of domestic producers of medicines is estimated at more than 20% in value terms, and 57% in physical terms at . At present, the enterprises of the

Belbiopharm Concern produce more than 600 products, 500 of which are medicines.

The main goal of the pharmaceutical industry of the Republic of Belarus is to produce medicines of various pharmaceutical groups and biotechnological products. The state policy in the field of drug supply is focused on import substitution and production of generics - analogues of well-known foreign medicines. Specialists note that the development and testing of an original drug requires at least 6 years and USD 100-200 thousand. Foreign companies invest much larger sums - up to USD 2 billion in the USA and up to EUR 70 million in Europe - in developing an absolutely innovative chemical substance and bringing it to pharmacies. By producing generics, millions of dollars can be saved, as manufacturers no longer spend money on drug development and testing, so the market in many countries is oriented towards generics. For example, the total value of their sales in the UK, Denmark and the Netherlands on the prescription drug market is 50-75% of all sales; the share of generics in Canada is 85%■ in the USA and Germany - from 20 to 45%.

It is quite difficult to develop a large range of original medicines in Belarus, as this requires huge financial investments, which, according to analytical estimates, will start to pay off only in 10-11 years or more. Currently, Belarus cannot afford such financial costs, so the development of original medicines is not a priority at the moment. This is also evidenced by statistical data. The share of original drugs in the Belarusian pharmaceutical market is 30%, while the share of reproduced (generics) drugs is 70%. Thus, pharmaceutical production in Belarus is oriented towards generics.

After the 20-year patent expires, any pharmaceutical company can synthesise the original compound or buy it from chemical companies. The copied drug is also subjected to trials, but with a shortened programme. The costs of generic production are lower, so the price is lower, but this does not mean that the quality is inferior. It is proved by tests of chemical identity and binoequivalence.

The policy pursued by Belbiopharm Concern provides for the creation of conditions for the development and production of high-quality and competitive products. In the pharmaceutical industry, this is achieved by certification of quality management systems for compliance with the requirements of international standards ISO 9000 and Good Manufacturing Practice (GMP). Quality management systems in accordance with the requirements of ISO 9000 have been implemented at nine pharmaceutical plants. National QMS certificates have been obtained by the Bourne-Grodia

Medical Preparations Plant, Belmedpreparaty RUE, and Exon RUE. The production sites at JV "Mniskiitercaps" are fully certified. More than 100 items of medicines are produced according to GMP norms (which implies full quality control at all stages of production), 11 production sites have been certified.

In order to reduce the share of imports in domestic pharmaceutical products, the State Programme for the Development of Import Substitution of Pharmaceutical Substances, Prepared Medicines and Diagnostic Products for 2010-2014 and until 2020 was adopted, which includes the following subprogrammes: "Pharmaceutical Substances and Prepared Medicines", "Phytopreparations", "Diagnostics", and "Personnel Training for the Chemical and Pharmaceutical Industry". The implementation of this programme will require financial support in the amount of 600 billion Belarusian rubles. Most of the resources (about 84%) will be spent on the modernisation of Belarusian enterprises and their certification according to theCMP standard. At the same time, the programme does not focus on the creation of original Belarusian drugs, but aims to create high-quality generics that will not be inferior to foreign analogues.

Thus, "economic priorities for the development of the Belarusian pharmaceutical industry include: expanding the range of medicines through import substitution policy; ensuring their high quality, safety, and effectiveness in accordance with international standards; increasing exports, stimulating the production of highly effective medicines; introducing import-substituting technologies; organising import-substituting in-house production of pharmaceutical substances within the framework of the State Programme of Pharmaceutical Substances in the framework of the State Programme of Pharmaceutical Substance Substitution.

Chapter 3

SUBJECT AND OBJECTIVES OF PHARMACEUTICAL BIOTECHNOLOGY

Pharmacology (Greek: phormocon- medicine, *logos*- teaching) is a science that studies the conditions, processes and consequences of the effects of drugs and other biological compounds on living organisms. Pharmacology as an academic discipline consists of two sections: general pharmacology and private pharmaco- ^{log}t ....

General *pharmacology* characterises the fundamental laws of action of drugs on the body and the general concepts of pharmacodynamics and pharmacokinetics.

Pharmacodynamics studies localisation (places and points of application) and mechanisms, types, types and effects of action of drugs; dependence of their action on various factors, e.g. chemical structure, dosage form, route of administration, dose, as well as on age, sex, body weight, general condition of the patient.

Pharmacokinetics determines the routes of administration, the processes of absorption of drugs by different routes of administration, transport and distribution in the body (biotransformation, metabolism) and excretion from it

Private pharmacology studies pharmacodynamics and pharmacokinetics of specific drugs, indications and contraindications for use, dosing and dispensing.Using data from biomedical disciplines, pharmacology as one of the basic sciences of drugs allows you to make the most complete conceptual understanding of them, to justify the creation of a rational dosage form, to determine the relationship between the chemical structure and the effect of drugs on the body.

Challenges of pharmaceutical biotechnology:

. production of highly effective pharmaceuticals used in healthcare and other sectors of the economy;

- finding new drugs for the prevention and treatment of diseases;

... study:

- mechanisms and effects of drug action;
- peculiarities of their organism;
- distribution in organs and tissues, metabo- lismal reactions and pathways of elimination;

. development of highly effective drugs for pro prevention and treatment of diseases in order to increase life expectancy and the period of labour activity of people.

Modern drugs can control neuro-psychic activity, blood circulation, respiration, metabolism, digestion and other functions of the organism. Drugs in uncontrolled "irrational use can cause undesirable, apori "very severe disorders of vital functions of the human body, so pharmacology, creating effective and safe drugs, simultaneously develops recommendations for their rational use Pharmacology, using the achievements of biology, microbiology, physiology, chemistry, medicine, pharmacy and other sciences, helped to solve the problems of treatment of most of the serious infectious diseases that have caused the death of millions of people. Nowadays it is possible to treat some previously considered incurable mental illnesses, diabetes mellitus, myocardial infarction, certain malignant neoplasms and a number of othersToday it is impossible to imagine the treatment of any disease without the use of drugs.

3.1. Sources of medicinal substances

In the arsenal of medicines, in addition to synthetic preparations, a significant place is occupied by preparations and individual substances of "medicinal raw materials (plant, animal and mineral). In this way, many widely used drugs are obtained in the form of not only more or less purified preparations (galenovs, novogalenovs, organopreparations), but also in- 28 / .

For example, the alkaloids morphine codeine and papaverine are isolated from opium; reserpine from the snake plant Rauwolfia reserpine; cardiac glycosides digitoxin and digoxin from foxglove; and hormones from a number of endocrine glands.

Some medicinal substances are products of fungi and microorganisms. Of these, antibiotics are of most interest.

It should be emphasised once again that medicinal substances of plant, animal, microbial and fungal origin often serve as the basis for their synthesis, as well as for subsequent chemical modifications and the production of semi-synthetic and synthetic drugs.

Alkaloids (al-qali (Arabic) - *alkali, eidos* (Greek) - species) are nitrogenous organic compounds found mainly in plants. Free alkaloids are bases. They are usually found in the form of salts in plants. Many alkaloids have high biological activity (morphine, atropine, pilocarpine, nicotine, etc.).

Glycosides are a group of organic compounds of plant origin, which are broken down by enzymes or acids into a sugar, or glycone *(gtykys* (Greek) - sweet), and a non-sugar part, or aglycone. A number of glycosides are used as medicines (strophanthin, digoxin, etc.).

3.2. The main ways of searching for new drugs, their

clinical trials

The progress of pharmacology is characterised by the continuous search for and creation of new, more advanced drugs. Their path from a chemical compound to a medicinal product can be presented in the following form:

. chem lab;

. pharmacological laboratory;

- laboratory for finished dosage forms;
- pharmacological committee;

. clinical trials;

- New Drug Introduction Authority;
- chemical-pharmaceutical industry;

. implementation in medical practice.

Drug development begins with the research of chemists and pharmacologists. The search for new drugs develops along the following lines.

1. ***Chemical synthesis of drugs:***

a) directed synthesis:

- nutrient reproduction;

. creating antimetabolites;

- modification of molecules of compounds with known biological activity;
- combination of the structures of two compounds with the required properties;
- synthesis based on the study of chemical transformations of substances in organisms;

б) the empirical way:

- chance finds;

. screening.

2. ***Preparation of preparations from medicinal raw materials and isolation of individual substances^ :***

. of animal origin;

- of plant origin;

. of minerals.

3. ***The isolation of drug substances that are products of fungi imilroorganisms.*** Currently, drugs are obtained mainly by chemical synthesis. One of the important ways of directed synthesis is the *reproduction of* biogenic *substances* formed in living organisms. For example, adrenaline, noradrenaline, γ-aminobutyric acid, prostaglandins, a number of hormones and other physiologically active compounds have been synthesised.

The *search for antimetabolites* (antagonists of natural metabolites) has also led to the development of new drugs. The principle of creating antimetabolites is to synthesise structural analogues of natural metabolites with opposite metabolite action. For example, antibacterial agents sulfonamides are similar in structure to para-aminobenzoic acid, which is necessary for microorganisms vital activity, and are its antimetabolites. By changing the structure of fragments of the acetylcholine molecule, it is also possible to obtain its antagonists.

The most common way of finding new drugs is chemical *modification of compounds with known biological activity.* The main objective of such research is to create new drugs that favourably differ from the already known ones (more active, less toxic). The starting compounds can be natural substances of plant or animal origin, as well as synthetic ones. For example, on the basis of hydrocortisone, produced by the adrenal cortex, many more active glucocorticoids have been synthesised that affect water-salt metabolism to a lesser extent than their prototype.

Recently, the possibility of creating new drugs *based on the study of their chemical transformations in the body has* attracted attention. These studies are developing in two directions. The first is related to the study of products of drug metabolism. In some cases, the resulting metabolites have a more pronounced activity than the original compounds. For example, from aitide-pressant imisin (imipramine) in the body is formed more active desmethylimipramine (desipramine). The latter is also used in as a medicinal product.

The second direction involves the study of mechanisms of chemical transformations of substances. Knowledge of enzymatic processes that secure metabolism of substances allows to create drugs that change the activity of enzymes. For example, acetylcholinesterase inhibitors (proserine and other anticholinesterase agents) have been synthesised, which enhance and prolong the action of the natural mediator acetylcholine. Inhibitors of the enzyme monoamine oxidase involved in the inactivation of noradrenaline, dopamine, and serotonin have also been obtained (these include the antidepressant Inalamnd and others). There are known substances that induce (enhance) the synthesis of enzymes involved in the detoxification of chemical compounds (e.g., phenobarbital).

In addition to directed synthesis, the empirical way of obtaining drugs still retains a certain importance. A number of drugs were introduced into medical practice as a result of *accidental discoveries*. For example, the decrease in blood sugar levels found with sulphonamides led to the synthesis of their derivatives with pronounced hypoglycaemic properties - butamnda and aialogical preparations used in the treatment of diabetes mellitus. The effect of teturam (Antabuse), used in the treatment of alcoholism, was also discovered accidentally in connection with its use in industrial rubber production.

One variation of empirical search is screening. In this case, any chemical compounds that may be intended for non-medical purposes are screened for biological activity using a variety of techniques. Screening is a very labour-intensive and inefficient way of empirical drug discovery. However, it is sometimes unavoidable, especially if a new class of chemical compounds is being investigated, the properties of which are difficult to predict based on their structure.

In pharmacological research, the pharmacodynamics of substances are studied in detail: their specific activity, duration of effect, mechanism and localisation of action. An important aspect of the study is the pharmacokinetics of substances (absorption, distribution and transformation in the body, as well as excretion routes). Special attention is paid to side effects, toxicity in single and prolonged administration, teratogenicity, carcinogenicity, mutagenicity. It is necessary to compare new substances with new drugs of the same groups. Physiological, biochemical, biophysical, morphological and other methods of analysis are used in the pharmacological evaluation of compounds.

Of great importance is the study of the efficacy of substances in appropriate pathological conditions (experimental pharmacotherapy). For example, the therapeutic effect of antimicrobial substances is tested on animals infected with certain infectious agents, antiblastoma agents - on animals with experimental and spontaneous tumours. In addition, it is desirable to have information about the peculiarities of the action of substances on the foyer of those pathological conditions in which they can be used (for example, atherosclerosis, myocarditis, inflammation). This direction, as already mentioned, has been called pathological pharmacology. Unfortunately, existing experimental models rarely fully correspond to what is observed in the clinic. Nevertheless, they simulate to some extent the conditions under which drugs are administered and thus bring experimental pharmacology closer to practical medicine.

The results of the research on substances promising as pharmaceuticals are submitted for pharmacological testing. If the conducted experimental pharmacological studies are exhaustive, the proposed compound is transferred to clinics that have the necessary experience in drug research.

The decisive role in the evaluation of new drugs belongs to clinical pharmacologists, whose main task is the clinical study of pharmacokinetics and pharmacodynamics of drugs, including new drugs, and the development on this basis the most effective and harmless methods of their use.

Clinical trials of new medicines should be based on a number of principles. First of all, they should be tested on a wide contingent of patients. In many countries, this is often preceded by a trial in healthy people (volunteers). It is very important that each new substance be compared with well-known drugs of the same group (e.g. narcotic analgesics with morphine, cardiac glycosides with strophanthine and foxglove glycosides). A new medicine must necessarily differ from existing medicines for the better

In clinical trials of substances, it is necessary to use objective methods that allow quantitative assessment of the observed effects. A comprehensive study using a large set of adequate methods is another requirement for clinical trials of pharmacological substances.

In cases where an element of suggestion can play a significant role in the effectiveness of substances, *placebos* are used. Placebos are dosage forms, which in appearance, odour, taste and other properties mimic the drug being taken, but do not contain the drug substance. Placebos consist only of indifferent form-forming substances

In so-called blind control, the patient is given alternating portions of drug and placebo in an unknown sequence. Only the attending physician knows when the patient is taking either or both. In double-blind testing, another person (the head of the department or another physician) is informed. This principle of substance testing allows a particularly objective assessment of the effects of substances, as placebos can have a positive effect in a significant proportion of patients in a number of pathological conditions (e.g. some pains).

The validity of data obtained by different methods should be confirmed statistically.

Only when a substance has obvious advantages over similar drugs or represents a new group of drugs can it be recommended for chem-

The programme has been developed and implemented in the micro-pharmaceutical industry and subsequently in the broader medical practice.

The quality of drugs produced by the chemical-pharmaceutical industry is usually assessed using chemical and physicochemical methods specified in the State Pharmacopoeia of the Republic of Belarus (hereinafter referred to as the State Pharmacopoeia). In some cases, if the structure of active substances is unknown and chemical methods are not sensitive enough, the so-called biological standardisation is used. It means determination of the activity of drugs on biological objects (on the basis of more typical effects). Hormone preparations, cardiac glycosides, etc. are evaluated in this way. Activity is expressed in conditional units of action (UD). A standard with constant activity is used for comparison. Methods of biological standardisation and substances for which they are mandatory are specified in the State Pharmacopoeia .

Chapter 4

REGULATORY SUPPORT FOR PHARMACEUTICALS BIOTECHNOLOGICAL INDUSTRIES

Regulatory documents supporting pharmaceutical biotechnological production:

- production specifications;
- technological regulations of production (laboratory, pilot, start-up, industrial, standard);
- pharmacopoeial articles.

4.1. Technical specifications for medicinal products

Technical conditions for medicinal products - a set of requirements to the characteristics of medicinal products, regulating qualitative and quantitative indicators that allow standardisation, certification, declaration (for the right to enter the foreign market). Example: TU 4311-182-38576343-92, where 4311 is the code of the product group according to the All-Belarusian Classifier of Products (OKP), 182 is the product code, 38576343 is the code of the enterprise according to the All-Belarusian Classifier of Enterprises and Organisations (OKPO), 92 is the year of TU development.

Specifications for medicinal products shall consist of:

- from the technical requirements;

. safety requirements;

- environmental protection requirements;

. the rules of acceptance;

- control methods;
- transport and storage conditions;

. manufacturer's warranties.

The introductory part should contain the name of the medicinal product, its purpose, field of application, conditions (rules) of use.

4.2. Technological regulations of production

Technological regulations are the main technological

is a document, part of the organisational-administrative and organisational-technological documentation systems of the enterprise, as well as the quality management system, therefore, when developing it, the compliance between the enterprise documents, authorisation documentation and registration dossier is not violated, as well as the requirements of supervisory authorities, regulatory and normative-legal documents that apply to the production of medicinal products are taken into account.

Technological regulations for the production of medicines are used as *a basic technological document:*

- when carrying out technological processes in serial production. ,
- development of initial data for the design or reconstruction of industrial production;
- Establishing technical and economic standards, including standards for raw materials and supplies;
- developing technological instructions, as well as instructions on safety, industrial sanitation and anti-fatigue measures;

. development and implementation of measures for utilisation of industrial waste, neutralisation and treatment of industrial effluents and discharge into the atmosphere.

Depending on the purpose, technological regulations are subdivided: ,

- ^laboratory
- pilot;
- launchers;

- industrial.

Technological regulations are developed and approved for each type (form) of medicinal product.

Laboratory regulations are the *first* normative technological document, the development of which completes scientific research in laboratory conditions to create a new drug (new technology).

The laboratory regulations establish methods of product manufacture and conditions to ensure reproducibility of technological processes under laboratory conditions with stable yields, as well as safety regulations.

Laboratory regulations are developed according to the same rules and contain the same sections as industrial regulations.

The parameters of technological processes and operations of manufacturing a new product (new technology) described in the laboratory regulations, as well as methods (techniques) of control, monitoring of critical points, terms and conditions of storage of the product are determined in the process of research work on the basis of planned comparative, controlled studies with mathematical processing of the results.

The laboratory regulations combine the requirements of Good Laboratory Practice (GLP) and Good Manufacturing Practice (GMP), as these regulations are used to prepare pilot (experimental) series of the drug under laboratory conditions, intended for preclinical studies and clinical trials on animals.

A brief feasibility study (or business plan) is an annex to the laboratory regulations.

In accordance with the "Industry standard of industrial regulations for the production of chemical-pharmaceuticals" OST 59.01.002.40-85 laboratory regulations for the production of antibiotics should include certain sections.

Characterisation of medicinal products. Includes the name of the antibiotic, main purpose, brief description of the drug properties, description of the organism forming the antibiotic, methods of determining biological activity, storage conditions.

Technological *scheme of production.* Reflects the sequence of work on antibiotic production with subdivision into stages. The process flow diagram serves as a basis for the future technology of industrial production of the drug.

Raw *materials.* The quality requirements of the raw materials used in the production of the antibiotic are reported in order to maximise yields and ensure repeatability. It is necessary to focus on raw materials and materials produced by the domestic industry.

Apparatus scheme of production. Presenting the process of antibiotic production with indication of apparatuses and devices, their design, size and other characteristics that may be important in antibiotic production.

Process flow statement. This is a description of the process of antibiotic production based on completed experimental results performed under laboratory conditions. The process is included in the regulations if reproducible results can be obtained both in terms of antibiotic quality and yield.

Production waste, process and ventilation emissions into the atmosphere, their utilisation and neutralisation. A list of possible wastes and atmospheric emissions, a list of valuable substances in wastes and recommendations for their utilisation, as well as substances harmful from the point of view of environmental pollution and methods of their neutralisation are given.

Production control. Specific requirements for the equipment are specified (tightness of the fermenter and all communications, serviceability and reliability of the agitator, etc.). The quality of raw materials is analysed for compliance with certain standards. Regimes of sterilisation of media and individual substances are determined. Methods of analysing the control over the process of antibiotic biosynthesis and finished products are selected.

Fire safety and industrial hygiene. A list of substances capable of ignition and explosion is given. All substances used in the antibiotic production process must be studied from the point of view of safety, fire hazard and industrial hygiene.

List of production instructions. All instructions to be developed on the basis of laboratory re- gla~ .... are listed. .

Technical and economic standards. The yields of final and intermediate products are determined; specific consumption rates of raw materials, specific consumption rates of steam, water, electricity, air.

Information materials. Biological and physicochemical properties of the substance, pharmacological properties (advantages and peculiarities) are indicated. Comparison with the indicators of identical foreign preparations is carried out, information on the patent purity of the antibiotic and the adopted method of its obtaining with a list of copyright certificates (patents), information on the harmfulness of substances used in obtaining the drug and precautions when working with them.

Op^otromkaillenntreglovement - a technological document that *completes scientific research* in the development of a new drug (new technology) in laboratory conditions.

Pilot industrial regulations, like industrial regulations, should contain sections covering all aspects of the technological process.

The basis for the development of pilot industrial regulations are:),

. laboratory regulations,

- results of research and development, preclinical and clinical trials.

Experimental and industrial regulations for new products shall contain:

- *Zoned for the design of a pilot industrial technology - a pilot line of a given capacity,* control, measuring and testing equipment used in the technological process;
- *technical and economic indicators* to clarify and supplement the feasibility study (business plan);
- *data from toxicological-hygienic studies* to justify hygienic regulations.

Experimental-industrial regulations for new products are developed simultaneously with the pharmacopoeial article or other regulatory document for the drug.

In case of organisation of pilot production (e.g. at the base of the enterprise-developer), for the production of small batches of new medicinal products for a long period of time, it is allowed to use pilot industrial regulations as the main technological document with its revision every three years and approval in the order established for industrial regulations.

A start-up* (temporary) *regulation is a technological document on the basis of which the industrial production of a newly created LS is developed.

The start-up regulations are developed by the enterprise on the basis of:

. pilot industrial regulations;

- design documentation for new production or on the basis of existing

production facilities, if their technology is fundamentally changed.
The start-up regulations must fully comply with the requirements for industrial regulations.
The start-up procedure is amended and supplemented as production progresses.
Production development is considered *complete* when:
- requirements of regulatory documentation "products;
- achieved the design data both in terms of capacity and key technical and economic indicators of production;
- the start-up regulations are replaced by industrial production regulations.

Further serial production of the medicinal product is carried out on the basis of industrial regulations.
The period of validity of start-up regulations in production is not more than three years.
Industrial regulations. Serial production of commercial products is carried out on the basis of industrial regulations, which are drawn up after the completion of the period of mastering the industrial production of a newly created drug instead of start-up regulations.
Industrial regulations include:
- characterisation of the finished product;
- product name;

- the category and number of the applicable regulatory document;
. registration number;
- information on the organisation (legal entity or individual) - manufacturer or supplier;
- the main purpose of the product and its consumer properties;
- safety conditions for use, storage, transport and disposal;
- packaging and labelling requirements;
- shelf life;
- production flow chart;
- general process flow diagram;
- diagrams of the stages and operations;
- plan of the production shop (site);
- hardware scheme of production and specification of equipment (this section of the regulations consists of a drawing of the hardware scheme of production and specification of equipment (Table 5) assigned to this particular production);
- characterisation of raw materials, intermediate products, starting and packaging materials (quality requirements for raw materials, materials and

intermediate products in this production);

- detailed description of production strains of microorganisms (in the production of immunobiological drugs):

- name and designation of medicinal products;
-location;
-historical excerpts;
-describing the methodology of attenuation;
- information on immunogenicity, pathogenicity, toxicity;
- full characterisation of morphological and other properties;
- conditions of propagation and seed preparation,
- conditions and terms of storage' transport and preservation methods _ .

. statement of the technological process (data on types and quantities of raw materials, materials, intermediate products used and obtained in production, waste, allowable losses and output of the finished product);

- description of technological works (sanitary preparation of personnel, premises and equipment; inspection and preparation of equipment for work; preparation and loading of raw materials; conduct and control of technological works; unloading and transfer for further processing of products (intermediate products, waste); packaging, labelling and shipping of finished products);
- material balance (complete information on the theoretical yield at each stage of production, indicating the maximum and minimum yields above WHICH.an investigation is required in accordance with company procedures). Acceptability limits of 95 to $^{10}5$ per cent of the nominal quantity for active substances and limits of 90 to 110 per cent of the nominal quantity for excipients do not require further justification;
- information on recycling and neutralisation of production waste:

- waste treatment processes (WTP);
-Waste Disposal (WD);
-technological (ventilation) emissions to the atmosphere,

... production control - a list of production points, the inspection of which ensures reliable compliance with the established regime of technological process (includes the place (code of stage or operation of technological process), the object of control, the name of the determined parameter and its standard, methods and means of control). Examples of mandatory control points:

- requirements for equipment and premises in preparation for work (sterility, tightness, integrity of anticorrosion coating, etc.);
- control of raw materials before they are used in the production process;

- control of basic parameters and requirements for loading raw materials and carrying out individual technological operations;
- quality control of the intermediate products obtained;
- control of indicators regulating the discharge of industrial effluents, etc;

. safe operation of production facilities. The fundamental documents are the standards included in *the system of labour safety standards. (SSSS).* Particular attention is paid to compliance:

- sanitary and epidemic regime;
- veterinary and sanitary regime;
- industrial hygiene standards;
- biosecurity requirements.

Table 5

Equipment symbols used in the drawings

Equipment	Designation
Measuring instruments	SI
Control systems	SR
Reactors, autoclaves, fermenters, etc.	P
Filter equipment	Ф
Drying plants	SU
Grinding and milling equipment	M
Tanks (collectors, sumps)	E
Compressors	К
Pumps	н
Refrigeration units	X
Filling and packaging equipment	У
Equipment	Designation
Thermostats	T
Centrifuges	ц

According to the requirements, labour safety *when working with biological objects* is ensured:

. production processes;

- production equipment;

. means of defence;

- a system of special preventive measures;
- environmental protection (list all emissions into the environment (dust, gaseous, liquid, solid), their name, source of emission (apparatus, stage), quantity with permissible deviations, frequency, duration, chemical composition, physical parameters). Indicate that the ventilation and air purification system ensures protection against emissions, infectious material and harmful substances into the environment. Indicate the measures to ensure

reliable protection of water resources and air basin in case of emergencies and production shutdowns for repair. Indicate the availability of emergency and drainage tanks, trays, flares, adsorbents, disinfectant solutions to prevent volley discharges into the environment;

- list of production instructions. In this section of the regulations allocate designations (codes) and names of instructions mandatory for the conduct of the technological process, including instructions:
- for all jobs in accordance with the staffing table[s] (job descriptions);
- The Company's employees are responsible for safety, industrial sanitation, veterinary, sanitary and fire safety;
- Preparation of equipment for work, repair and reception from re[montai;], ...
-
- operation of equipment, measuring and automation equipment;
- disaster response;
- prevention of contamination of raw materials and finished products during storage and production;
- packing and packaging;
- The care and maintenance of production and control animals;
- technical and economic standards. Here you can find *cost standards* characterising the technical level of production and technological process according to the requirements of the regulations, including:
- norms of consumption of basic raw materials and supplies,
- norms of technological costs (steam, water, electricity, liquid air, inert gas);
- labour input per unit of final product;

. information materials - information:

- the developers of the drug and its production regulations, including the date of development;
- experimental works of the enterprise and the organisation, confirming the correctness of the choice of optimal parameters of technological processes;
- the hazards of the substances used in production and the precautions to be taken when working with them;
- the patent purity of the product and the method of preparation adopted, listing the patents protecting the product;
- quality indicators of foreign analogues of similar products, which have similarity in functional value and conditions of application;
- literature sources and reports on research, development and design works, which are referred to in the text of the regulation; for manuscripts, the places of their storage are indicated.

4.3 Pharmacopoeial article

Pharmacopoeial article (PA) is a normative-technical document approved by the relevant authorised executive authority and having the character of a state standard, in which the requirements to the quality of a medicinal product or medicinal herbal raw material, its packaging, conditions and shelf life, quality control methods are specified.

Pharmacopoeial article includes the name of medicinal plant raw materials both in Russian and in Latin (the Latin name fulfils the function of the international name).

Types of pharmacopoeial articles:

1. ***Pharmacopoeia Article of the enterprise (PPA)*** is a quality standard of a medicinal product under the trade name. It contains a list of methods and indicators of quality control of a medicinal product, taking into account the specific technology of the enterprise, which has been examined and registered in a certain order in accordance with the State Pharmacopoeia, and quality indicators should not be lower than the requirements contained in the State Pharmacopoeia.

The term of validity of the pharmacopoeial article (pharmaceutical article) of the enterprise is established at its approval *for no more than five years,* taking into account the level of technological process of a particular production of a medicinal product.

2. ***General, pharmacopoeial article. (GPC)*** is a state quality standard for a medicinal product that contains the basic requirements for the dosage form, as well as a description of standard methods of quality control of medicinal products.

OFSincludes:

- a list of standardised indicators and test methods for a particular dosage form;
- description of chemical, physical, physico-chemical, biological, biochemical, microbiological methods of analysis of medicines;
- requirements for titrated solutions, reagents, indicators used

Pharmacopoeial article, general pharmacopoeial article and pharmacopoeial article of the enterprise are approved by *the head of the enterprise* with obligatory assignment of designation.

3. ***A private pharmacopoeial article (PPP)*** is a normative and technical document regulating the quality and safety of a medicinal product. It is created for a medicinal product under the international nonproprietary name (if available) or under the name that replaces it in the mandatory established procedure. It also includes a list of normative indicators and test methods for the given medicinal product, as well as references to general pharmacopoeial articles.

*4. **Temporary pharmacopoeial article*** is a normative-technical document approved for the period of development of industrial release of a medicinal product and for the development of industrial technology of methods for determining the quality or indicators of a new medicinal product for a period not exceeding three years.

Structure of the pharmacopoeial article:

- *introductory part :*
- time of collection of raw materials (vegetation phase, sometimes calendar period), obligatory characterisation of raw materials according to the mode of their technological processing: dried, threshed, freshly harvested, freshly frozen, etc;
- A wild or cultivated plant;

-egojnzneinform;

- the name of the producing plant and family in Russian;
- *external signs :*

-composition of raw materials;

- characteristic diagnostic features, odour and taste (for non-venomous species), size of raw material;

-microscopy ■

- diagnostic signs of the anatomical structure of raw materials (luminescence microscopy is given for some species);
- the type of microdrug in which the study is performed;

. qualitative.reactionsSprnrn are conducted^ qualitative.hy- stochemical reactions and chromatographic authenticity samples for the main groups of active substances, the methodology of their performance and results);

- *numerical indicators :*

-for whole, cut or powdered raw materials (the indicators are a standard for all types of medicinal plant raw materials and determine their quality);

- content of active extractive substances, total ash and ash insoluble in 10% hydrochloric acid solution, impurities and grindability;

- *Quantification* (describes the methodology for quantifying the total content of all active substances in terms of any substance contained in the raw material);
- *packaging,* (types of packaging and weight of raw materials per unit of packaging are specified);
- *microbiological purity* (specifying the method of determination of microorganisms and their permissible limits);

- *labelling* (given in accordance with the requirements for graphic design of medicinal products);
- *transport* (if necessary, requirements for loading, unloading and handling of products after transport are specified);
- storage (defines the conditions of product storage, including requirements for product protection from climatic factors);
- *shelf* life (the time during which the medicinal raw material can be used);
- *pharmacological action* (pharmacological group to which the medicinal raw material is classified).

Chapter 5

GENERAL CHARACTERISATION BIOTECHNOLOGICAL PROCESSES

Biotechnology is the use of cell cultures of bacteria, yeasts, animals and plants, whose metabolism and biosynthetic capabilities ensure the production of specific substances. Biotechnology emerged at the junction of several biotechnological sciences, such as genetics, bacteriology, virology, molecular biology, microbiology, biochemistry, plant science. An important role was played by the discovery of ways to modify DNA and its transfer from one organism to another (Fig. 6).

Figure 6. Linking biotechnology with other sciences

Historically, biotechnology originated from traditional microbiological industries such as bread, cheese, wine, beer and dairy products. These technologies are still of great importance and are constantly evolving.

Biotechnological processes are carried out by using bacteria, yeasts, mould fungi, algae, cultures of cells and tissues of plants and animals. These processes utilise the metabolic peculiarities and biosynthetic capabilities of cells. The goal may be the production of cell biomass or products of cell activity - metabolites.

Currently, biotechnology is divided into industrial and applied microbiology, genetic and cell engineering. The main directions of industrial microbiology are presented in Table 6.

Table 6

Main directions of biotechnology in various industries

Industry	Scope of application
Agriculture	Production of protein and vitamin concentrates. Breeding, cloning and genetic engineering of animals and plants. Use of antibiotics for treatment of animals and birds. Production of vaccines. Production of bioinsecticides. Use of hormones and other growth stimulants
Production of chemicals and compounds	Production of organic acids. Production of vitamins, antibiotics, etc. Use of enzymes in the composition of synthetic detergents (SDS)

Environmental monitoring	Improved methods of testing and monitoring of environmental contaminants. Use of microorganisms to treat agricultural, domestic and industrial waste
Medicine	Application of enzymes in diagnostics and microorganisms in the creation and modnfication of complex drugs. Synthesis of new antibiotics, hormones and interferons. Use of enzymes and microorganism strains in medical practice
Energy	Biogas and ethanol production
Materials science	Ore leaching. Study and control of biodegradation
Food industry	Creation of new methods of processing and storage of food products. Use of food additives obtained with the help of microorganisms. Utilisation of unicellular proteins. Application of enzymes. Improvement of alcoholic and lactic acid fermentation

The need for biotechnology is driven by shortages of food, energy, mineral resources and the need to improve health and environmental protection.

Bioindustry includes sectors in which biotechnology can replace widely used traditional methods and sectors in which it has always played a leading role (see Table 6).

Rapid returns occur in the following biotechnological ot- (races)∧ yah: _

1) improving digestion;

2)biogas production;

3) production of safe and affordable vaccines;

4) Bioenergetics;

5) improving composting techniques;

6) cellulose hydrolysis;

7) increasing nitrogen fixation by symbionts.

Only a few hundred species out of more than 100,000 known microorganisms are used in industry, as an industrial strain must fulfil a number of strict requirements:

. grow on cheap substrates;

- have a high growth rate or produce a high yield in a short time;
- synthetic activity towards the desired product; the formation of by-products should be low;
- to be stable in terms of productivity and to the requirements of the cultivation conditions;
- be resistant to phage and other types of infection;
- be harmless to people and the environment;
- Thermophilic, acidophilic (or alkophilic) strains are desirable as they are

easier to maintain sterility in production;

- anaerobic strains are of interest, as aerobic strains create difficulties in cultivation - they require aeration;
- the resulting product must have economic value and be easily distinguishable.

In practice, strains from four groups of microorganisms are used·

- yeast;
- mycelial fungi (moulds);
- bacteria;
- ascomycetes.

The term yeast, strictly speaking, has no taxonomic meaning. They are unicellular eukaryotes belonging to three classes: *Ascomycet.es, Basidiomycetes, Deuteromycetes.*

Ascomycetes include primarily *Saccharomyces cerevisiae*, certain strains of which are used in brewing, winemaking, bread making, and ztyl alcohol production.

The ascomycete *Saccharomyces lipolytica* metabolises nephthyl hydrocarbons used for protein production.

The denteromycete *Candida utdis* as a source of protein and vitamins is grown on non-food raw materials: sulphite liquors, hydrolysates of wood and liquid hydrocarbons.

The denteromycete *Tnchosporon cutaneurn* oxidises many organic compounds, including toxic ones (e.g. phenol), and is used in waste water treatment.

Mycelial fungi use:

. in production:

- organic acids: citric acid (*Aspergillus niger*), gluconic acid (*Aspergillus niger*), itaconic acid (*Aspergillus terreas*), furmaric acid *(Rhlzopuschrysogenurn);*

-antibiotics (penicillinicephallosporin);

- speciality cheeses: *camembert* (*Pemcillium camamberti), Roquefort (PenicUUUumroqueforti);*

- hydrolysis process in solid media: in rice starch to produce sake, in soya bean to produce tempeh, miso.

The beneficial bacteria are related to the denticellular bacteria.

Lactic acid bacteria of the genera *Lactobacillus, Leuconostoc, Lactococcus* have been used *industrially* since ancient times.

Acetic acid bacteria of the genera *Acetobater, Gluconobacter* convert ethanol into acetic acid.

Bacteria of the genus *Bacillus* are used in the production of toxins harmful to insects, as well as for the synthesis of antibiotics and amino acids. Bacteria of the genus *Corynebacterium are used* for the production of amino acids.

Among actinomycetes, the most common are the genera *Streptomyces* and *Micromonospora, which* are used as antibiotic producers. When growing on solid media, actinomycetes form a thin mycelium with aerial hyphae that differentiate into chains of coidiospores.

The following compounds are currently being synthesised using microorganisms:

. alkaloids;

- amino acids;
- antibiotics;
- antimetabolites;

antioxidants;
proteins;
vitamins;
herbicides;
enzyme inhibitors;
insecticides;
ionophores;
coenzymes;
lipnds;
nucleic acids;
nucleotides and nucleosides;
oxidising agents;
organic acids;
pigments;
surfactants;
polysaccharides;
anthelmintics;
antitumour agents;
solvents;
plant growth hormones;
sugars;
sterols and converted substances;
iron transport factors;
pharmacological agents;
enzymes;
emulsifiers.

5.1. Classification of pharmaceutical manufacturing products

Biotechnological production is based on the use of microorganisms. In order to control the microbiological process, it is necessary to know the physiology of the microorganism cultures used. This will allow to control the processes occurring in the cell, the conditions of cultivation and the influence of the main environmental factors on the directed biosynthesis.

The products of biotechnological production are natural macromolecules - proteins, enzymes, polysaccharides, polyesters, isolated from the cells of microorganisms, tissues and organs of plants and animals.

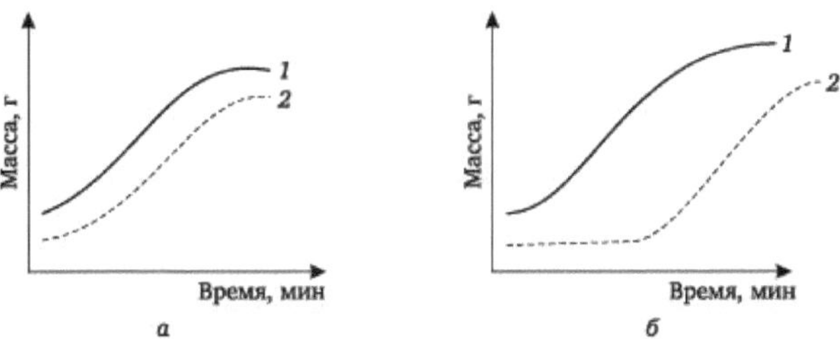

Figure 7. Dynamics of biomass change and formation of primary (*a*) and secondary (b) metabolites during organism growth:
1 - biomass; *2* - product

In relation to growth processes.low molecular weight metabolic products of living cells are divided into primary and secondary metabolites (Fig.7).

Primary metabolites are low molecular weight compounds (molecular weight less than 1500 Da) necessary for microbial growth. Some of them are building blocks of macromolecules, others are involved in the synthesis of coenzymes. Among the most important metabolites for industry are amino acids, organic acids, purine and pyrimidine nucleotides, solvents and vitamins

Secondary metabolites are low molecular weight compounds that are not required for growth in pure culture. Secondary metabolites include antibiotics, alkaloids, plant growth hormones and toxins.

5.2. General biotechnological scheme of pharmaceutical production

The processes of biotechnological^ production are diverse, but all have five basic stages:

- preparation of nutrient medium;
- microbial culturing;
- obtaining seed;

- isolation of the target product;

. purification of the target product.

The preparation of pitotelic media is a stage that fulfils two basic requirements that must be met by the Wo-nerve medium: it must be nutritionally complete and inexpensive. Carbon and nitrogen in a digestible form are necessary for protein biosynthesis; phosphorus for the synthesis of DNAiadenosine triphosphoric acid. (ATP); trace elements. - For the formation of enzymes, growth factors and vitamins are also needed for normal life activity. Secondly, the medium must be sterile, which is achieved by temperature, ultraviolet, ultrasonic and ultrasonic treatments.

Aultivation (fermentation) is a set of sequential operations from the introduction of inoculum into a pre-prepared nutrient medium to the termination of growth and biosynthesis processes due to the depletion of nutrients in the medium. There are two main types of fermentations: biomass production of micro-organisms and metabolite production.

Seed (inoculum) is *obtained* according to the scheme shown in Fig. 8:

Figure 8. Scheme of seed production

The quality of the obtained seed is monitored by microscopy.

Pharmaceuticals require a *high degree of purity.* The cost of purification is higher the lower the concentration of the substance in the cells. Purification steps:

. separation;

- destruction of cell membranes (biomass disintegration);

. cell-wall separation;

. separation and purification of the product;

- fine purification and separation of preparations.

5.3. Separation

Separation - separation of the mass of the producer from the liquid phase. Preliminary separation may be carried out to increase efficiency:

- pH change;
- heating;
- addition of protein coagulants or flocculants.

Methods.of.separation:

1. (literally, surface water swimming) - section-
introduction of small particles and introduction of dispersed phase droplets from the fluid - based on different wettability of particles (droplets) with liquid (mainly water) and their selective adhesion to the interface surface, usually "liquid-gas" (very rarely "solid-liquid").

The main types of flotation:

- _ foaming (culture liquid with microorganisms' biomass is continuously foamed by air supplied from bottom to top under pressure, cells and their agglomerates "stick" to finely dispersed air bubbles and float with them, collecting in a special settling tank);
. oil;
. film.

2. *Filtration is the* process of percolation, the natural filter*ing* of liquid through porous substances. Filters are used in this process:

. of single and multiple uses;
• batch and continuous operation (with automatic removal of pore-clogging biomass layer);
• drums;
. discs;
. belts;
. cymbals;
. carousel vacuum filters;
• filter presses of various designs;
. membrane.

3. *Physical sedimentation is a* process in which biomass is precipitated by the addition of lime or other solid components that entrain cells or mycelium to the bottom.

4. *Centrifugation is a* method in which suspended particles are precipitated by centrifugal force to form two fractions: biomass (solid) and culture liquid. This method requires expensive equipment, but allows for maximum particle freeing of the culture liquid.

Centrifugation and filtration can be carried out simultaneously in filtration centrifuges.

High-speed centrifugation separates cellular components by size (larger particles move faster during centrifugation) and is used if the products sought are within the cells of the producer.

5.4. Destruction of cell membranes (biomass disintegration)

Disintegration methods:

- mechanical (physical);

. chemical;

-combined

Physical methods-ultrasound treatment, rotation of a paddle or vibrator, shaking with glass beads, squeezing through a narrow pressure hole, crushing frozen cell mass, grinding in a mortar, osmotic shock, freeze-thaw, decompression (compression followed by a sudden release of pressure). These methods are economical, but their use may reduce the quality of the product .

Chemical and chemical-enzymatic methods - cells are destroyed by toluene or butanol, antibiotics, enzymes, are more selective.Examples:

- cells of Gram-negative bacteria are treated with lysozyme in the presence of ethylenediaminetetraacetic acid or other determinants.
- yeast cells - snail zymoliazoa, enzymes of fungi, ac- tinomycetes.

5.5. Separation and cleaning of products

The isolation of the target product from the culture fluid or homogenate of destroyed cells is carried out by precipitation (precipitation), extraction or adsorption.

Precipitation :

- physical (heating, cooling, dilution, concentration.
- chemical (using inorganic and organic substances - ethanol, methanol, acetone, isopropanol).

Mechanism of deposition. by organic substances, reduction of the dielectric constant of the medium, destruction of the hydrate layer of molecules- .

The dissociating ions of inorganic salts are hydrated during *salting*.

Reagents: ammonium sulphate, sodium sulphate, magnesium sulphate, phosphate (kalyl)-.

Extraction is the process of selective extraction of one or more soluble components from solids and solutions by means of a liquid solvent, the extractant. Types of extraction:

. solid-liquid (a substance from the solid phase changes into the liquid phase: e.g. chlorophyll from an alcohol extract into petrol.

. liquid-liquid (substance passes from one liquid to another: extraction of antibiotics, vitamins, carotenoids, lipids.

Extragents: phenol, benzyl alcohol, chloroform, liquid propanilibutanidr.

Ways to improve the efficiency of extraction:

- re-extraction with fresh extractant;

. selection of the optimum solvent;

- heating of the extraction agent or the liquid to be extracted.

. lowering the pressure in the extraction apparatus.

Extraction with chloroform in laboratory conditions is carried out in a "Soxhlet" apparatus, which makes it possible to repeatedly use the dis-taoritel.

Adsorption, a special case of extraction (when the extracting agent is a solid), follows the ion-exchange method.

Adsorbents are cellulose-based ionites:

. cationite - carboxymethylcellulose (CMC)

- anionite - diethylaminoethylcellulose (DEAE);
- dextran-based sephadexes, etc.

5.6. Methods for the fine purification of pharmaceuticals

Chromatography, (Greek *chroma*, genitive case of

chromatos - colour, dye) - a physico-chemical method of separation and analysis of mixtures, based on the distribution of their components between two phases - stationary and mobile (eluent), flowing through the stationary phase.

Types of chromatography according to the techniques of chromatography:

- *Column - the* separation of substances is carried out in special *columns*;
-
- Thin layer (*TLC)* - separation takes place in a thin layer of sor[benta;], . .

. paper - on special paper.

For large scale separation and purification of biotechnological process products are applicable:

Affinity precipitation *is a* process in which a ligand is attached to a soluble carrier and, upon addition of a mixture containing the corresponding protein, a complex is formed with the ligand, which precipitates immediately after its formation or after addition of electrolyte to the solution;

- *affinity separation is the* most highly efficient of the affinity purification methods, based on the use of a system containing two water-soluble polymers;
- hydrophobic *chromatography is a* technique that involves the binding of a protein by the interaction between the aliphatic chain of an adsorbent and the corresponding hydrophobic site on the surface of the [protein (1b|;)].
- *electrophoresis is a* method of separating proteins and nucleic acids in free water, solution and porous matrix, which can be polysaccharides such as starch or agarose. According to the principle of separation electrophoresis can be:

-natnwy;

- SDS (in the presence of sodium dodecyl sulphate);
- two-dimensional, etc.

Chapter 6

BASIC PRINCIPLES OF PROTEIN PRODUCTION IN THE PHARMACEUTICAL INDUSTRY

6.1. Use of microorganisms for protein production

In accordance with the nutritional standards, a person should receive from 60 to 120 g of complete protein daily.

The constant synthesis of various protein compounds is necessary to maintain vital functions of the organism, to build cells and tissues. While most plants and microorganisms are able to synthesise all amino acids from carbon dioxide, water, ammonia and mineral salts, humans and animals cannot synthesise some amino acids (valine, leucine, isoleucine, lysine, methio-NII, threonine, tryptophan and phenylalanine). These amino acids are called essential and must be supplied with food. Their deficiency causes severe diseases in humans and reduces the productivity of farm animals.

At present, the world deficit of protein is about 15 million tonnes. Microbiological synthesis is the most promising. While for cattle it takes 2 months to double protein mass, for pigs - 1.5 months, for chickens - 1 month, for bacteria and yeast - from 1 to 6 hours. The world production of food protein products due to microbial synthesis is more than 15 thousand tonnes per year.

Let us consider an example: the doubling time of the Escherichia coli is 20 min, i.e. in 20 min two daughter cells will be formed from a single cell, in 40 min - four "granddaughters", in 60 min - eight "great-granddaughters", in 80 min - 16 "great-granddaughters". In 10 h 40 min from one bacterium more than 6 billion will be formed, and in 44 h from one bacterium with a mass of 1 - 10-^{12} g a biomass of -stvub - Yu^g will be formed.

The use of various microorganisms as sources of protein and vitamins is due to the following factors:

... the possibility of using a variety of chemical compounds, including industrial waste, for cultivation of microorganisms.

. relatively uncomplicated microbial production technology that can be carried out year-round and automated.

- high content of protein (up to 60-70 %) and vitamins, as well as carbohydrates, lipids in microbial preparations;
- The protein has a higher content of essential amino acids compared to plant proteins;

. the possibility of directed genetic influence on the chemical composition of microorganisms in order to improve the protein and vitamin value of the

product.
Industrial production of food products based on microorganisms requires thorough biomedical research. Such products should undergo comprehensive testing to detect carcinogenic, mutagenic, embryotropic effects on human and animal organisms. Toxicological studies, digestibility of microbial synthesis products are the main criteria for the feasibility of their production technology
Proteins are produced by yeasts, bacteria, algae and nmncellular fungi
The advantage of yeast over other microorganisms lies in their processability: resistance to infection, ease of separation from the medium due to the large size of the cells. They are able to accumulate up to 60 % of protein rich in lysine, threonine, valine and leucine (these amino acids are scarce in plant feeds). The mass fraction of nucleic acids is up to 10%, which is harmful to the organism.^ As a result of their hydrolysis is formed mio-go purine bases, which are then transformed into uric acid and its salts, which are the cause of urolithiasis, osteo- ochoidrosis and other diseases..The optimal rate of yeast additives in the feed of farm animals is from 5 to 10% of dry matter. Yeast is used for food and feed purposes.
The advantages of the bacterium are its high growth rate and ability to synthesise up to 80% of protein, which contains many deficient amino acids: methionine and cysteine. The disadvantages are small cell size and low concentration in the culture medium, which makes isolation difficult. Some bacterial lipids may contain toxins. The mass fraction of nucleic acids is up to 16 %. They are used only as feed.
The advantages of algae are significant content of protein of full amino acid composition, accumulating in the amount of 65 %, easy isolation of algae from the culture medium, low content of nucleic acids - 4 % (for comparison: in higher plants 1-2 %). Algae are used for food and feed purposes.
Mycelial mushrooms are traditionally used as a food product in African countries, India, Indonesia, China, etc. They accumulate up to 50 % of protein, with an amino acid composition approaching that of animal protein, and are rich in B vitamins. Cell walls are thin and easily digested in the gastrointestinal tract of animals. The mass fraction of nucleic acids is 2.5 %.
Since 1985, microbiological protein has been used in the food industry for the manufacture of various products and semi-finished products.
Three main forms of microbial protein are used in food production:
. whole, mass (without breaking down cell walls);
• partially purified biomass (involves breaking down cell walls and removing unwanted components);

.isolated isobiomass proteins (isolates)

The World Health Organisation (WHO) has concluded that microbial protein can be used in food, but the permissible amount of nucleic acids introduced together with it into the diet of an adult should not exceed 2 g per day. The introduction of microbial protein does not cause negative consequences, but there are allergic reactions, gastric diseases, etc.

6.2. Use of yeast for protein production

The technology of baker's yeast production was developed in Germany in the 19th century, during the First World War yeast was used as a food additive in the production of soups and sausages, and the technology of fodder yeast production began to develop.

Until now, the cultivation of *Saccharomyces serene (carlsbergensis)* brewer's yeast remains an important reserve of dietary protein and vitamins. The human body assimilates more than 90 % of all nutrients contained in them. There are 14 vitamins found in this yeast, and it is particularly rich in B vitamins.

In the processing of biomass in. food whitecake is thoroughly purified. First, the yeast cell walls are destroyed by mechanical, alkaline, acidic or enzymatic treatment, followed by extraction of the homogeneous yeast mass with a suitable organic solvent. The proteins are then dissolved with an alkaline solution and the protein solution is separated from the cell mass by dialysis. Proteins purified from low-molecular impurities are precipitated and used as protein additives in various food products - sausages, sausages, pates, meat fillings. The dry protein can also be texturised.

Some yeast cells (genera *Candida, Rhoäotorula, Torulops*, Tnchosporon*) are able to use unbranched hydrocarbons with a number from 10 to 30 carbon atoms in the molecule as a carbon source for growth. They are mainly represented by liquid fractions of petroleum hydrocarbons with a boiling point of 200...320 ^{O}C. Initially, the project originated from the need to utilise paraffins remaining in the amount of 10 to 15 % after gas oil refining. Macro- and microelements, vitamins and amino acids are added to the nutrient medium. In Russia and the SFSR a plant for production of fodder yeast on oil paraffins was built in 1971 (its productivity was about 1 million tonnes per year). The dried protein mass is pelleted and used as a protein-vitamin concentrate in fodder production.

Milk whey is a good substrate for the cultivation of fodder yeasts of the genera *Torula, Kluyveromyces*. About 10kg of protein and 50kg of lactose are contained in 1t of whey. Protein is separated by ultrafiltration and the lactose solution is used for yeast culture.

Yeast cells can also use lower alcohols (methanol and ethanol) obtained from natural gas or plant waste as carbon sources. In this case, yeast mass contains more proteins (56-62 % of dry weight) and less harmful impurities (benzene derivatives, D-amino acids, abnormal lipids, toxins, carcinogens) than fodder yeast grown on petroleum paraffins.

Yeasts grown on plant.hydrolysates of *Candida arborea* and *Candida utilis* are used for food purposes and as protein supplements to various products. For example, in the United States, *Candida utilis* is used to produce torutei, which is added to foods and is then considered a high protein dietary food.

6.3. Utilisation of bacteria

More than 30 bacterial species are known that can be used as sources of complete feed protein.

Natural and associated gases, hydrogen, and alcohols such as methanol, ethanol, and propanol can serve as a source of carbon for bacterial culture.

The most commonly grown bacteria of the genus *Methylococcus, which* can utilise 85 to 90 % of methane in special fermenters, are grown on gaseous media. However, production of feed protein on gaseous media is rather expensive. The technology of bacteria cultivation on methanol, which is easily obtained by methane oxidation, is more widely used. Most often bacteria *of Methylomonas, Methylophilus, Pseudomonas* genera are used. A number of enterprises produce a fodder preparation. This preparation contains up to 74% protein (of dry matter), up to 5% lipids, 10% mineral substances, 10 to 13% nucleic acids.

Bacteria with high intensity of protein synthesis include hydrogen-oxidising bacteria capable of accumulating up to 80 % protein (per dry matter). For their cultivation the gas environment should contain from 70 to 80 % hydrogen, from 20 to 30 % oxygen, from 3 to 5 % carbon dioxide. Production can be organised near chemical plants.

6.4. Use of algae

As early as 1521, after the conquest of Mexico, it was reported that the Aztecs ate cheese-like cakes. In Lake Chad (Africa), the natives of the Kanembu tribe eat balls of blue-green algae.

The protein content of Chlorella and Sceiidesmus cells is about 55 % (in terms of dry matter), and in Spirulina cells - 65 %. Algae are well balanced in amino acid composition (except for methiotin), they contain quite a lot of polyunsaturated fatty acids ß-carotene.

No abnormal pathological effects were detected when feeding Ispirulina to animals, and the growth rate was normal.

Protein mass from algal cells is marketed as a suspension, dry powder or paste preparation. The process of separating the algae cells from the water mass is the most labour-intensive stage.

6.5. Use of mushrooms

Fungi of the genus *Rhizopus* sp. are used for solid-phase fermentation of soya beans. After three days, the mycelia of the fungus grows and binds the beans into a crust containing up to 40% protein. In Indonesia, this crust is fried and added to soups as a meat substitute. This product is called tempeh. Various grain legumes are fermented in a similar way in Africa and the East.

The advantage of solid-phase fermentation is lower energy costs, the disadvantage is low productivity.

Various mycelial fungi are grown on starch (grain) materials, citrus peels, straw, bran, and husks. They are used either for protein enrichment of fodder or for isolation of enzymes from the resulting biomass.

The fungus *Fusarium graminearum* is cultivated to produce the food product mycoprotein. It is grown on cheap glucose syrup made by hydrolysis of wheat or maize starch. Mycoprotei is well digested. It is given a coney flavour and aroma of meat, ham, fish. The product retains its flavour for a long time and does not shrink during cooking. Industrially produced in Great Britain.

6.6. Protein purification methods

6.6.1. Preparation of the extract

Starting material. The choice of organisms of one or another species used for the isolation of proteins and enzymes is based mainly on the ease of their cultivation. Of animals, the most widely used is the rat (especially in liver work)

and rabbit (when working with skeletal muscles). If work is carried out with organs that are too small in laboratory animals (heart, brain, kidneys, thymus), they are taken mainly from bovines and pigs, since the study of enzymes from human tissues is limited for obvious reasons. On the other hand, human blood is readily available and there are many methods for isolating enzymes from red blood cells and other blood cells. Blood plasma is poor in enzymes but contains significant amounts of non-enzymatic proteins. Recently, the study of enzymes from various species has become more routine. This is partly due to the realisation that many important features of metabolic control, enzyme localisation and other characteristics vary considerably within vertebrate species.

Little attention is paid to invertebrate farm proteins, mainly because most invertebrate animals are very small and many individuals have to be used to obtain the required amount of starting material. If the desired organ is not extracted from each individual (a difficult task when working with insects), the purification of homogenate prepared from a whole body is complicated by the fact that it contains a large amount of digestive juice with proteolytic enzymes, which are also released into the extract. Most of the well known invertebrate enzymes are derived from large crustaceans such as river crayfish and crab.

For a long time, plant biochemistry has had very little connection with animal biochemistry, mainly because of the medical orientation of the latter, but also because the enzymologist faces certain difficulties when working with plants. Qualitative differences due to seasonality can be overcome if plants are grown in special chambers; otherwise, the variability of open-air material must be tolerated. For obvious reasons, out of the great variety of species, those of economic importance are used for study, but from the biochemist's point of view, this is not always the best choice. Nevertheless, one or two plant species are predominantly used because extracts can be prepared from the leaves without great difficulty and chloroplasts can be easily isolated from them. *Spinach Spmacia olerácea* and the somewhat related beetroot *Beta vulgaris*, which grow well in a variety of climates, are the most commonly used. Plant cells have many compartments; in most cases the bulk of the cell is occupied by a vacuole filled with highly acidic solutions, proteases, and a variety of other substances harmful to the cytoplasm. A significant part of the cytoplasm is occupied by chloroplasts, starch granules and other organelles; the cell also has a voluminous cell wall (cellulose). The cytoplasm (which, together with chloroplasts, contains most of the enzymes) may often comprise no more than 1-2 % of the total volume of the plant cell. Consequently, plant extracts contain very little protein, even if a minimal amount of liquid is added during their preparation

When working with microorganisms, other problems arise. Animal material can be obtained from the local abattoir or from an institute vivarium, plant material can be obtained from the market, and microorganisms (with the exception of yeasts) need to be specially cultivated under specific conditions and in rather large quantities. Each group of microorganisms such as algae, fungi, yeasts and bacteria requires special conditions for cultivation, cell collection and extraction. A particular problem requiring careful study is the significant change in enzyme composition during the different phases of cell growth. It is desirable to collect bacterial cells and other unicellular microorganisms during the logarithmic phase of growth, although the activity of the enzyme to be studied may not be maximised during this period. Preliminary studies of the organism should be carried out to determine the

physiological state at which the desired enzyme is most abundant
Baker's yeast (*Saccharomyces cerere*) is available in large quantities, so if it can be used, the raw material problem is solved. Commercially available pressed yeast consists of almost 100 % yeast cells at the end of the growth phase. Yeast is grown on various nutrients. The main source of carbon for them is molasses or some other source of sucrose. At 0 C, yeast remains viable (and thus the proteins retain their nativity) for several weeks, and when stored frozen it can be used after several months. Dry yeast, whose viability is restored by adding water to it, is very popular in the baking industry and with biochemists. They serve as a very convenient source of many enzymes, although the activity of some of the most sensitive enzymes may be slightly reduced during drying. When stored in a vacuum or nitrogen atmosphere at 0...4 ^{O}C, they can be used even after several years.
It is very important to ensure that the desired protein or enzyme present in the starting material is extracted from it completely so that its content in the tissue can be reliably determined. If cell disruption is incomplete, the object may be underestimated as a source of enzymes because much of the enzymatic activity is not released into solution but is removed with the precipitate during extraction. If the enzyme is isolated from bacterial material, the experimenter must consider the possibility of selecting strains with abnormally high levels of enzymatic activity. This can be done either by conventional mutant selection or by using a more sophisticated cloning technique whereby multiple copies of the enzyme gene are incorporated into the bacterial genome using plasmids. Expression of the gene can be further enhanced by fusing it to a strong promoter. This work is time-consuming and can only be done by an experienced technician; it is not worth even attempting such methods if there is little time available for enzyme curing.
Freshness of raw material and its storage. Usually, the sooner the raw material is used, the better, i.e. the preparation is more consistent with normal physiological state, because if the tissue has been stored for a long time, natural degradation processes begin. However, there are cases when absolute freshness can be a hindrance. For example, skeletal muscles contain a large amount of ATP in the presence of phosphocreatine and glycogen high concentration of ATP is maintained for several hours and after death. Structural proteins myosin and actin are normally insoluble at ionic strengths below 0.25, but in the presence of ATP, ruptured myofibrils swell and part of the proteins pass into solution. Even at physiological ionic strength (-0.16), some amount of myosin can pass into the soluble state. Subsequently, this may affect the quality of enzyme preparations, as myosin precipitates and

mixes with fractions in which it should not be present. An example of a special kind is the preparation of adenosine monophosphate-deaza-minase. When extracting very fresh muscle with buffer with a recommended ionic strength of 0.26, a lot of actin and myosin are transferred into the solution, which prevents the next step - adsorption on phosphocellulose. A better quality preparation can be obtained if the muscle is first allowed to enter a state of rigor mortis, which is accompanied by a loss of ATP. This is a rare occurrence. It is usual to work with fresh material.

The opportunity to obtain fresh material is always matched by the opportunity to use it. Both the obtained material and the extract should be stored frozen. Many events occur during the freezing process. First of all, free water freezes and ice crystals form. These destroy membranes and opra-nellae, but do not directly damage proteins and enzymes. As the temperature approaches the eutectic point for the various salts present in the solution, these salts crystallise. The least soluble salt is the first to precipitate out of the buffer pair, so the pH value can change dramatically before the solution is completely frozen. If stored at -15 to -25 ^{O}C (the usual temperature for evaporators in domestic refrigerators and simple freezers), the pH value of the concentrated solution that remains unfrozen will be different from the pH value of the original solution. In such a concentrated protein solution, proteases released from lysosomes during the freezing process may begin to function (albeit slowly at such low temperatures), and storage of the material for several weeks may severely compromise its quality. Thus, when freezing, the temperature should be reduced very quickly to -25 ^{O}C and the material should be stored, if possible, even at a lower temperature. Commercial freezers provide temperatures down to -80 °C.

Sometimes it is better to freeze the extract, as it can be changed to achieve optimum storage conditions. Once the extract is obtained, proteolytic enzyme inhibitors can be added and its pH can be adjusted with a suitable buffer to a value at which the biomolecules are stable. In addition, the extract can be obtained from large quantities of material, so that samples taken subsequently from the freezer will differ only in storage time, their properties in this case not being due to biological differences in the frozen material.

6.6.2. Cell disruption and extraction

Most of the proteins and enzymes studied in the early history of protein chemistry have been isolated from extracellular fluids. The reason for this is not only the ease of obtaining the material, but also the fact that extracellular proteins are generally more stable, often due to the presence of disulfide bridges, as well as the small size of the molecules. Initial studies of protein structure naturally focused on these small proteins. Thus, lysozyme, ribonuclease, and chymotrnpsin were the earliest proteins to be studied in detail, and all were derived from extracellular sources. However, most enzymes are localised within cells. These proteins are often less stable - they usually lack disulfide bonds due to the fact that the intracellular environment has reducing properties. The purpose of this section is to describe methods for cell disruption and isolation of enzymes into an aqueous extract, which is the first step in enzyme purification.

There are many methods of cell destruction due to the diversity of cell types. Most have inherent characteristics that must be taken into account when destroying them. Living tissues vary widely in their strength: in addition to the easily destroyed red blood cells, there is the tough collagen-containing material found in blood vessels and other tissues that contain smooth muscle. Plant cells are usually more difficult to break down than animal cells because they have a cellulose shell. Among bacteria, there are also quite fragile organisms that can be destroyed by digestive enzymes or simply by osmotic shock , and more resistant organisms with thick cell walls. To destroy the latter, a strong mechanical impact is necessary, which, however, should not be excessive, as it may inactivate labile enzymes.

Table 7 lists the techniques used to disrupt cells and extract enzymes into solution.If the enzyme is localised in organelles, the method can only be applied if the organelles are also disrupted. On the other hand, it may be necessary to pre-extract the organelles themselves, thereby getting rid of cytoplasmic enzyme contamination before extracting the enzyme from the organelles For. this requires a less severe exposure. Pre-dissolution of surface collagen and cellulose structures by preparations of hydrolytic enzymes allows further cell disruption by gentler methods, preserving the integrity of the organelles. An example of such a technique is the preparation of mitochondria from tissues such as skeletal or cardiac muscle under the action of proteolytic enzymes. However, this technique is only applicable for obtaining preparations in small quantities and is more suitable for metabolic studies of organelles than for enzyme purification. The yield of purified organelles can be very low, so in many cases it is better to first destroy the whole tissue and then proceed to isolate the desired enzyme from the complex mixture of proteins in the extract. Thus mitochondrial and chloroplast enzymes are often obtained from tissue homogenate rather than from isolated organelles.

Table 7

Methods of cell destruction

Method	Example	Principle
Soft impact		
Cell lysis	Erythrocytes	Osmotic disruption of the cell membrane
Enzyme-mediated degradation	Treatment of bacteria with lysozyme	Disruption of the cell wall leads to osmotic rupture of the cell membrane
Chemical solubilisation and autolysis	Yeast extraction with toluene	The cell wall (membrane) is partially dissolved by the action of chemical substances, the lytic enzymes that are released complete the process
Manual homogenisation	Liver tissue	Cells are squeezed through the narrow gap, resulting in the destruction of the cell membrane
Grinding (rubbing)	Muscle tissue, etc.	Cells are destroyed in the process of crushing the tissue by shear force
Impact of medium force		
Paddle homogeniser (Warringah type)	Muscle tissue, most animal tissue, plant tissue	There is mechanical destruction of large cells and separation of small cells from each other
Rubbing with abrasive (e.g. sand or aluminium oxide)	Plant tissue, bacteria	Cell walls are destroyed by the presence of micro-roughnesses on the abrasive particles
Method	Example	Principle
Strong impact		
French press	Bacteria, plant cells	The cells are forced through a small hole under very high pressure; they collapse under shear force
Ultrasound	Cell suspensions	Ultrasound waves create a high local pressure tra- dient; as a result, cells are destroyed by shear stress and cavitation
Ball mill	Cell suspensions	The cell wall is destroyed by the rapid vibration of the glass beads
Homogeniser Manton to Gowlin	Cell suspensions	Works the same way as a French press, but allows you to process more material

Finally, the enzyme may be insoluble in the buffer used for extraction.

After cell disruption and precipitation of insoluble material, an extract is obtained by centrifugation. The mixture before centrifugation is usually referred to as homogenate. After centrifugation, the liquid phase should contain as much of the enzyme present in the starting material as possible. Some of the liquid is retained by the precipitate, and the total loss will be proportional to the ratio of the volume of the precipitate to the volume of the liquid phase. In the case of extracts, losses due to retention of liquid by the precipitate should not exceed 20 per cent unless the raw material is very readily available and a high concentration of enzyme in a small volume of extract is more important than complete isolation of the enzyme from the homogenate. If the volume of liquid retained by the precipitate is half the volume of the precipitate, it can be calculated that for 80 % of the enzyme to be extracted, the volume of the supernatant must be twice as large as the volume of the precipitate.
volume of sediment. Most animal tissues contain a significant amount of insoluble cellular material, which binds a large volume of water. When dealing with them, the precipitate volume is almost equal to the original tissue volume (depending to some extent on the centrifugation conditions). Thus, when a homogenate is obtained, e.g. from liver, at least two volumes of suitable extraction buffer should be added. The more buffer added, the more of the soluble fraction will be extracted, but the extract will be more dilute; its large volume may cause difficulties in further work. For the preparation of homogeates from liver, heart and skeletal muscle, 2.5 volumes of extraction buffer are usually added.

Plant tissues differ significantly from animal tissues. As stated earlier, the intracellular content is only a small part of the volume of plant tissue. The presence of large vacuoles (considered here as extracellular formations) and intercellular spaces is the reason for the release of large amounts of fluid during tissue destruction. In this case, there is virtually no need to add extraction fluid. The volume of the precipitate after centrifugation may be only 20-40 % of the volume of the original plant tissue. Nevertheless, the use of additional extraction solution can be very important to control undesirable processes that occur during homogenisation. Such processes may be acidification of the medium or oxidation of unstable compounds. A particular problem is posed by plants containing phenolic compounds, which are oxidised mainly by endogenous phenoloxidases to form dark pigments. These pigments combine cov tively with proteins and inactivate many enzymes. Two ways of solving this problem are suitable. Firstly, the addition of a thiol compound, such as ß-mercaptoethayol, minimises the action of phenoloxidases. Secondly, the addition of powdered polyvinylpyrrolidone adsorbing phenolic compounds is often useful.

Microorganisms are more similar to animal tissues in terms of the volume of sediment obtained after centrifugation. Yeasts and similar fungi with thick cell walls produce sediment with a volume almost equal to the original volume of the cellular material. The same is true for bacteria. If significant shear forces are applied during the extraction process, a large amount of cell wall material may be transferred into solution.

Cell fragments can be centrifuged easily, especially large animal and plant tissue debris. However, after centrifugation, e.g. at 10,000 g for 15 min, intracellular particles remain in suspension. The turbidity of the resulting solution is not important as it can be removed during the fractionation steps. The use of more powerful cell disruption devices, such as a ball mill, makes it possible to break down cell walls to tiny particles of relatively low density, due to the fact that they usually contain lipids. Such particles are difficult to precipitate at the g values available for large volumes and the extract may remain very turbid. Methods for clarifying the extract are described below.

Animal tissue. Tissue is cut into pieces, connective tissue and fat are removed as far as possible. The cut tissue is placed in a paddle homogeniser, 2-3 volumes of cold extraction buffer per 1 g of tissue are added. The mixture is homogenised for 30 s; homogenisation is repeated if pieces of undestroyed tissue remain. When the homogenate is brought to the required value, it is stirred for 10-15 min and then transferred to centrifuge ^{1Sh1,}.T$^{(JAP1rY1}$)-C^{ents}R^{i}FU$^{(gi)}$RU$^{(ie}$) at 5000-10 000 g 5,000-10,000 g for bO min (2 - 10($^{5)}$-3 - 10$^{(5)}$g min). The extract is drained either through a *Miracloth* filter, or through an almarle or glass wool to remove fat particles.

Erythrocytes. The extract can be easily obtained from centrifuged erythrocytes after washing them with isotonic NaCl solution (0.9 %, 0.15 M) and re-centrifugation. The cells are destroyed by osmotic shock in water (2 volumes of water per 1 volume of centrifuged cells). However, about 90% of the protein that passes into solution is haemoglobin, and if you are denaturing another protein, it is very useful to use a method for selective removal of haemoglobin. An ethanol-chloroform mixture has been successfully used to denature haemoglobin.

Soft plant tissues. Crusty tissue is added only 0.5-1 volume of cold buffer containing 20-30 mM mercaptoethanol and homogenised for 30 s or passed through a domestic juicer prewashed with buffer. The homogenate should be centrifuged as soon as possible to reduce darkening due to oxidation (2 - 10^5-3 - IO^5 g - min). The liquid is carefully drained from the surface of the precipitate. The addition of powdered polyvinylpyrrolidone is useful for adsorption of phenols.

Yeast ! The cells can be completely destroyed with *a Manton-Haulin homogeniser,* using about two volumes of buffer per 1g of crude mass.

2. *Autolysis with toluene.* Various methods are used with toluene; however, not all of them are suitable for use with commercial yeasts. These methods are based on treatment of yeast with toluene, usually at a temperature of 35...40C. After 20-30 minutes, the yeast "liquefies" due to the extraction of cell wall components. After that, buffer is added to the yeast and the yeast is stirred for several hours or left overnight in the cold. Since the method is autolytic and cell wall structures are broken down by enzymes, some cell enzymes may be degraded during processing. Ethyl acetate can be used instead of toluene, but it is not suitable for yeast strains with myeloprosthetic cell walls.

3. *Ammonia cytolysis.* Eutmeyud is most suitable for working with dry yeast. It is simple, but some enzymes that are not stable at pH 10 may be completely lost. Active dry yeast is stirred in 0.5 M NH_4OH solution (2 volumes of solution per 1g dry matter) for 16-20 h at room temperature. The addition of a small amount of toluene (5-10 % of the total volume) sometimes helps to dissolve the proteins. 1-2 volumes of water and the necessary amount of acetic acid are then added to lower the pH to the desired value. Afterwards, the debris of the destroyed cells is removed by centrifugation.

Bacteria. Bacterial cells can be destroyed by ultrasound, with a ball mill or French press, although not all of these methods are convenient for preparing large volumes of extract. In the case of small volumes, cells can be rubbed with aluminium oxide. Gram-positive bacteria are usually sensitive to lysozyme. Stirring the cell suspension in buffer containing 0.2 mg/ml egg white lysozyme at 37 °C for 15 minutes is sufficient to release cytoplasmic components. Treatment of the extract with deoxyribonuclease I (10 μg/ml) improves its quality, as the viscosity of the liquid decreases.

Gram-negative bacteria are less sensitive to lysozyme if not pretreated. A method of combined treatment of cells with a non-ionic detergent, osmotic shock and lysozyme has recently been described. A modification of this method is used-

are used in the extraction of Gram-negative microorganisms *Zym.om.onas mobilis* and *Photobacterium phosphoreum.* In this case, exhaustive extraction of enzymes is achieved, although complete dispersion of the cell wall does not occur.

To 10 .- dry cell mass add 0.1 ml of 10% (by volume) solution of Triton X-100, 10 μl of mercaptoethanol and 2.5 ml of glycerol. The cells are dispersed and stirred vigorously for 30 min. Then 30 ml of extraction buffer (e.g. 20 mM KP pH 7.0₊ 1 mM ethylenediaminetetraacetate ODTA) containing 0.2 mg/ml lysozyme and10 μg/ml deoxyribonuclease is rapidly added and stirring is continued for 30 min. To the mixture 5 mg phenyl methylsulfonyl fluoride dissolved in 0.5 ml acetone and 0.1 mg pepstatin A were added, after which the suspension was centrifuged (15,000 g for 20 min).
These are just a few examples of procedures for extracts from various sources. In practice, it is more common to use the published procedure, but it should be borne in mind that the raw materials used as well as the equipment used may differ considerably from those described in the procedure, so that it is sometimes necessary to adapt the method to particular conditions.
Thus, extraction methods are based on crushing the material in an appropriate buffer followed by centrifugation of the mixture to remove insoluble residues.

6.6.3. Optimisation and clarification of the extract

The frequency behaviour of proteins during fractionation depends not only on their characteristics, but also on the composition of the solution, which also includes other proteins. Failures may be due to suboptimal homogenisation conditions (low pressure in the press cell, wear of the balls in the ball mill, dulling of the homogeniser blades), which leads to incomplete cell disruption. As a result, the extract may be more dilute and its composition may be different, as not all components are transferred into solution when cells are incompletely disrupted. To evaluate this extract, it is important to know the exact enzyme content per gramme of starting material. It is probably best to first obtain an extract from a small amount of starting material, using a large volume of extraction buffer (say 10 ml/g), and sequentially determine the enzyme activity at different time intervals. As the processing time increases, the enzyme activity.is likely to gradually decrease as the processing itself denatures the proteins either by heating or by shear forces during the grinding of the material. It is often necessary to cool the system during the cell disruption process. For this purpose, a chilled buffer is used and short periods of material processing alternate with pernodamnocooling. As mentioned earlier, it is very important to consider the volume of extraction buffer per gram of material. When using large quantities of starting material, a compromise has to be made between the desire to maximise extraction and the desire to obtain a minimum volume of extract.
For certain purposes, the extraction buffer should contain substances such as

ß-mercaptoethanol. Water can also be used for extraction, but it does not extract all proteins. It would be wonderful if the desired enzyme were completely extracted by water. Cells contain different suns and a lot of insoluble, charge-bearing material - proteins, phospholipids, nucleic acids. The ionic strength in the cytoplasm of a typical cell ranges from 0.15-0.2 M. Under these conditions, cytoplasmic proteins are soluble in the sense that they can move in the cell. When cells are homogenised with 2-3 volumes of water, the ionic strength of the homogenate may decrease to 0.05 and below. Under these conditions, charged particles in solution can act as ion exchangers and adsorb proteins, especially basic proteins.

If the homogenate is prepared with dilute buffer, the more buffer is added, the greater the fraction of protein that passes into the insoluble fraction. The ion-exchange adsorption properties (as opposed to physiologically important specific interactions) of cellular components are not always taken into account. In order to be sure that all soluble cell contents have been transferred to the extract, a buffer with an ionic strength close to the physiological strength and, of course, an appropriate pH value should be used. If too many undesirable components are extracted, it is better to go to a com- promise. Usually the following buffers are used: 20-50 mM phosphate, pH 7-7.5, 0.1M Tris-HCl, pH 7.5 and 0.1 M KC1 with a small content of buffer; for organelle extraction isoosmotic buffer mixtures containing sucrose, mannitol or sorbitol along with salts and buffer ß-mercaptoethanol ions are used. The buffer may also contain EDTA or cysteine (5-20 mM) and specific η.

stabilising agents for certain proteins, such as Zn for zinc-containing proteins or pyridoxal phosphate for enzymes that use it as a cofactor.

After cell disruption in the appropriate buffer, it is advisable to check the pH of the homogenate. Although after homogenisation of really fresh material in the buffer, the homogenate has the same pH as the buffer. This value may change due to metabolic processes leading to acidification of the medium. A good example is skeletal muscle, where glycogen is rapidly converted to lactic acid and the pH of the homogenate may drop from 7.0 to 6.0 in 30-60 min (including centrifugation time). In such cases, the pH value can be raised above the desired value before centrifugation (e.g. with 1 M Tris-buffer). In addition, the addition of a glycolysis inhibitor can stop the process. Fluoride (10- 30 mM) is sometimes used for this purpose.

These remarks relate to the improvement of extraction conditions for all soluble components. The extraction of a single enzyme or protein is better carried out in a medium in which all components are not completely solubilised and the desired object is extracted.

The extract is often cloudy. After centrifugation, fat particles may float on the surface. The fat can be removed by coarse filtration through glass wool or dense filter cloth. Insoluble particles of suspension contain organelles and membrane fragments which are only completely precipitated at 100,000 g, a procedure very inconvenient for large volumes of extract. Filtration is usually useless in such cases, for if the filter is dense enough to retain particles, it quickly becomes clogged. However, if turbidity is negligible but a clear extract is required, e.g. for passage through an adsorption column, filtration can be carried out with the aid of celite.

Different types of extracts. In some animal tissues, particles are a minor part of the extract and can be neglected. They aggregate at the first fractionation with ammonium sulphate and are then removed. On the other hand, there are animal tissues containing more fat and membrane structures; when extracting such tissues, suspensions are formed. Acidification of the medium to pH 6.0-5.0 usually results in aggregation of the material, which can then be centrifuged at a relatively low speed. Ribosomes and other nucleoproteins are usually removed by acidification, which can be regarded as a form of isoelectric precipitation, the phosphate groups being protonated and neutralising or at least reducing the charge of the suspension particles. This method gives very good results provided that the desired protein: a) does not precipitate isoelectrically at a given pH value; b) is not adsorbed on the resulting precipitate; and c) remains stable at non-physiological pH values. The extract is cooled and the pH is lowered with an appropriate acid (e.g. 1 M acetic acid). After stirring for 10-20 min, the precipitate is separated by centrifugation and the pH of the supernatant is adjusted, if necessary, to the desired value before proceeding to the first fractionation step. Plant tissues differ primarily in that the reaction of the medium is more acidic and the particles they contain, such as chloroplast fragments, are much more difficult to aggregate. In addition, plant extracts contain very little corpuscular material, apart from coarse particles such as starch grains, which easily precipitate after preparation of the extract.

A number of problems arise in the preparation of extracts from microorganisms. Firstly, many nucleic acids are extracted from rapidly proliferating cells. Secondly, in the process of cell destruction, the cell wall material is either highly dispersed, resulting in a turbid solution, or partially dissolved, and the solution contains many gum-like polysaccharides along with proteins and nucleic acids. This creates certain difficulties at the first stages of fractionation.

Extracts prepared from bacterial material by French press, ultrasound or lysozyme treatment are viscous due to the presence of DNA and contain significant amounts of ribosomal material. All nucleic acids precipitate as a result of their aggregation with macromolecular polycatoids. Salmon fish protamine and herring milk clupein are good reagents for this purpose, as binding to DNA is a natural function of these substances. The commonly used protamine sulphate should be used with great care: it has strong acidic properties and should be dissolved in water and neutralised before use. The amount of protamine sulphate required should be established. This can be as high as 5 mg per 1g of material to be treated. Protamine sulphate precipitates DNA, ribosomal nucleic acids, matrix RNA (mRNA), transport RNA (tRNA) and other nucleic acids. Some proteins can also adsorb on the resulting precipitate. Adsorption on protamine precipitate is even used as a step in the process of enzyme purification. If the use of protamine is not possible or desirable, the viscosity of the DNA solution can be reduced by using DNAase. Ribosomes can be precipitated with streptomycin. Although streptomycin is a poorer precipitator of nucleic acids than protamine, less protein is lost by ad-sorption to the forming precipitate.

After removal of nucleic acids, capsule gum-like carbohydrates may remain in solution. This creates difficulties in the use of conventional methods of protein precipitation. Ammonium sulphate and other fractional precipitation methods cannot be used in the presence of these substances. It is better to precipitate the protein completely so that the gum-like substances remain in solution. Virtually all of the protein can be precipitated at 80% saturation with ammonium sulphate and 55% concentration of acetone. If ammonium sulphate is used for protein precipitation, high-speed centrifugation may be necessary. Even then, the dissolved protein precipitate contains substances that reduce the reproducibility of results on the column or other fractionation methods. Two ways of overcoming these difficulties have recently been found. The first is amphiphilic adsorption on an agarose gel, in which all proteins are adsorbed on agarose pellets in a 70-80 % saturated ammonium sulphate solution, while non-protein material is washed out. When the agarose pellets are extracted with buffer, the protein passes into solution, which is then fractionated in the usual way. The second way is to use an appropriate extraction method. When treating cells with lysozyme, usually not too many capsule gum-like substances are not formed - lysozyme dissolves them, but does not destroy bacterial cells To obtain a sufficiently transparent extract from bacteria, the use of lysozyme together with a non-ionic detergent was successful. Once the nucleic acids have been removed, the extract can be used directly for fractionation. Such problems do not arise with all bacteria. Many studies have been published on the isolation of enzymes from bacterial material in which conventional fractionation following treatment of the extract with protamine or streptomycin has been successfully carried out.

There are several problems associated with the preparation of yeast extracts. The main one is that the cell walls of the fungi are very thick and difficult to break down, so more efficient methods are required to obtain the extract. If the extract is to be obtained quickly, the only reliable method is to use a vibrating ball mill or an Mznton-Haulin homogeniser. These methods produce a large number of low-density particles that are not completely removed by centrifugation. Turbidity may reduce the efficiency of protein precipitation by ammonium sulphate. However, it can be eliminated by aggregating the particles in an organic solvent (acetone, 20-25 % by volume at 0 °C). Most enzymes remain in the clear solution and are precipitated by increasing the concentration of acetone. Transparent extracts are also obtained using autolysis with toluene, but proteolytic cleavage of proteins may occur under these conditions.

Cytolysis under concentrated ammonia is used to isolate the very stable enzyme phosphoglucomutase. This procedure was developed and is widely used in "active dry" yeast. It is used to obtain clear extracts containing most of the glycolytic enzymes. Some enzymes are completely destroyed at high pH values. In this method, proteolysis is not a serious problem because the activity of proteases is quite low at pH 9.5-10. The extraction efficiency can vary considerably depending on the batch of yeast.

6.6.4. Methods for purification of proteins and enzymes associated with particles

A number of specific methods are widely used for the isolation of proteins and enzymes structurally associated with insoluble cell components such as mitochondria, chloroplasts, plasmalemma, endoplasmic reticulum, and nuclear membranes. These methods do not apply to water-soluble proteins enclosed within organelles, such as mitochondrial matrix proteins, but only to molecules covalently bound or firmly associated with particles. A significant degree of purification compared to the starting material can be achieved by repeatedly washing the homogenate with a buffer in which the protein in question is not soluble. Thereafter, two approaches are possible.One is to fractionate the precipitate by differential centrifugation (e.g., sedimentation or density gradient), by electrophoresis, or by "molecular sedimentation". ,
Another method is to solubilise proteins (subfractionation may precede

solubilisation). Not all enzymes are able to exist in a solubilised state isolated from their normal cellular environment; therefore, the success of purification of such compounds depends on the extent to which particle fragments containing the enzyme can be separated from other particulate material.

Particle-bound proteins can be solubilised by various $p_{(U)tions\cdot?}$o Many solubilised proteins behave subsequently like all water-soluble proteins and can be fractionated using the procedures described in Sects. 3-5. On the other hand, it is sometimes necessary to solubilise a substance using detergents. In this case, the detergent bound to the protein replaces the lipid-containing membrane. If the detergent is subsequently removed, the protein may denature, aggregate and drop out of solution.

Most of the membrane-bound proteins and enzymes studied in detail belong to the first category: they can be isolated by physical, chemical or enzymatic methods; the resulting solution is further fractionated in the usual way. Many such enzymes are found in mitochondria, including those associated with the electron transfer system. They are isolated by methods such as:

1) mitochondrial disruption for the isolation of matrix proteins followed by separation of membrane proteins from mitochondrial fragments using; a) ultrasound especially at high pH and temperature; b) alkali (pH 8-11) in the presence or absence of metal chelating agents (EDTA); c) solubilisation by detergents; d) organic solvents (e.g. acetone powder, extraction with n-butanol, ethanol, etc.); e) phosphate (e.g. acetone powder, extraction with n-butanol, ethanol, etc.). etc.); e) phospholipase A for membrane lipid cleavage;

2) lipid extraction and acetone powder production. The particle suspension is treated with several volumes of acetone at a very low temperature (e.g. -20 C or below). The dehydrated powder thus obtained can be stored for a long time. Proteins are dissolved by dispersing them in an aqueous buffer solution. This is a classical method that is still in use today.

Most membrane-bound proteins can be extracted from membranes in the presence of detergents, which include lipophilic chains that interact with the protein's hydrophobic uppers and displace it from the complex with the normal membrane. The most widely used detergents are sodium deoxycholate and Triton. Triton is the trade name of a whole series of mostly non-ionic detergents based on polyethylene glycol. Triton X-100 is most commonly used for a variety of purposes, although other types of detergents have some use. Detergents are capable of displacing proteins that are firmly bound to the membrane by hydrophobic interactions by firstly dissolving the membrane and secondly replacing the membrane components with aliphatic

or aromatic chains that form the lipophilic part of the detergent.
If the protein is solubilised and its integrity is found to be intact, fractionation can be performed. In most cases this is not a problem. The biological activity of the protein must be preserved; this depends on the choice of a suitable detergent. Nonionic detergents, such as Triton X-100, are very mild in their action, and most proteins, both membrane-bound and free, can tolerate concentrations of Triton X-100 as low as 1-3 % (w/w). In contrast, some anionic detergents (e.g. dodecyl sulphate) have, in addition to solubilising properties, a very strong denaturing action. They are not likely to be widely used in enzyme isolation.
Lipoproteins contain covalently or non-covalently bound lipid, which in turn is embedded in the membrane and serves as a link between the protein and the particle. Extraction in the presence of a detergent can result in the removal of the lipid and its replacement by the detergent. This may cause loss of enzymatic activity, which is restored by the addition of a natural lipid, usually phosphatide. Such proteins are particularly difficult to work with as they are very sensitive to the ratio of detergent to lipid in solution. Removal of the detergent invariably leads to aggregation and most often to deiaturation.
Excess detergent can interfere with fractionation. For example, salting with ammonium sulphate leads to the appearance of a layer of Triton X-100 on the surface of the solution, which often contains the desired proteins. However, effective separation does not occur. Column chromatography can be performed or proteins can be separated by gel filtration, but it is possible that the detergent micelles will move in the same zone as the protein and therefore end up in the same fraction. Ion-exchange chromatography has been successfully carried out in the presence of non-ionic detergents. Indeed, Triton X-100 at a concentration of up to 1% has a negligible effect on the ion-exchange properties of normal water-soluble proteins. But solubilised membrane proteins can only be in the composition of detergent micelles, ^ significantly affects the process of ion exchange. If the protein under study can be adsorbed on the nonoo exchanger, the excess detergent passes freely through the column, which allows elution of the protein free (otiosee quently) from the detergent. On the other hand, if complete removal of the detergent results in "denaturation of the protein, a small amount of detergent (< 0.1 %) is added to the buffer to prevent this. The collected fraction will, of course, also contain some detergent. However, since a particular enzyme is usually isolated from the protein mixture, the presence of a small concentration of pure detergent, uncontaminated by fat, in the final preparation will do little

harm.

Membrane-bound proteins differ in solubility from typical cytoplasmic ethical enzymes, and some unusual empirically found methods are sometimes used (and quite successfully) to purify them.

6.7. Biotechnology of interferon production

In 1957, E. Isaacs and A. Lindeman discovered that the cells of animals infected with viruses secrete into the sphere a factor that is able to cause resistance to viral infection in hentactyly (not treated with virus) cells . This factor was called *interferon.*

Interferon is a protein or glncoprotein with 146-166 amino acid residues. The molecular weight is 20-80 kDa.

Effects of interferon:

1) has *antiviral* action - stimulates the formation of interleukin-2, increases the synthesis of enzymes (endonucleases are able to "cut" the nucleic acid molecules of viruses at the level of translation);

2) causes *inhibition of cell growth* (used as an *antitumour agent*), is able to suppress the division of oncogenic cells while maintaining the function of activation of all parts of the immune system;

3) has a *stimulating effect on phagocytosis,* natural *killer* cells *and macrophages; increases non-specific cell resistance;*

4) has *antimicrobial activity;*

5) has a radioprotective effect, -

6) is an *immunomodulator*, i.e. it regulates immunological reactions and stimulates the body's immune system.

Interferon is produced by vertebrate cells.

The most active interferon producers are *lymphocytes* and *macrophages*

The most active *inducers, among viruses, are* Newcastle disease virus, Sendai virus, and swine fever virus.

Interferon classes:

D) leukocytic, or."*-IFN, is* obtained from cultures of leukocytes isolated from the blood of donors. There are 20 recombinant variants differing in the sequence of amino acids in the polypeptide chain and in their biological activity;

2) *fibroblast,* or *ß-IFN, is* isolated from fibroblast culture;

3) *immune*, or *u-IFN*, synthesised by sensitised *T-lymphocytes* upon repeated contact with mutagens, as well as with bacterial and viral antigens (AG).

All three classes of interferons have different physicochemical properties and differ serologically.

There are the following types of interferonisation:

- *exogenous* - introduction of ready-made interferon into the body, use of ointments containing interferon;
- *endogenous* - introduction of interferon inducers into the body, which stimulate the process of interferon formation.

The most active inducers of interferon are *synthetic and natural double-stranded RNAs.*

The mechanism of action of inducers.gRA is as follows:

- stimulate phagocytosis and antibody biosynthesis;
- inhibit tumour growth and metastasis;
- have anti-cellular activity;
- have a radioprotective effect;

- increase the sensitivity of cells to the action of interferon;
- are involved in the regulation of protein biosynthesis in the cell.

Step-by-step biotechnology for the production of GRA:

. isolation of informative RNA (nRNA), which carries information about the structure of the interferon molecule (RNA is isolated after induction of interferon synthesis);

. production of complementary DNA (cDNA) on matrixiRNA;

- cDNA insertion into the vector plasmid using restriction enzymes and ligases and production of recombinant DNA (rDNA) containing vector DNA with genetic markers and DNA encoding the synthesis of the target product;
- rDNA transformation into a recipient (permissive) microbe NuYuCellCu;
- obtaining clones of recombinant producers capable of synthesising interferon (on dense nutrient medium);
- propagation of a clonal culture - producer in liquid nutrient medium;
- separation of cells from the culture liquid by centrifugation ,
- isolation of target proteins from native solution or lysate of the producer biomass, e.g. by ammonium sulphate precipitation;

. purification. of interferon after dissolution of the precipitate by affinity chromatography using a sorbent bound to interferon monoclonal antibodies.

When a solution containing the target product and ballast substances is passed through a column with an immunosorbent, highly specific binding of the antigen (interferon) occurs. The interferon is then eluted with weak acid and the antigen-antibody complex dissociates.

To obtain bomctleukocyte interferon, 100 litres of blood or 1 litre of culture fluid of the recombinant microorganism are needed.

There are n other ways to produce interferon:

- based on *engineered recombinant DNAs* expressed in *E. coli* cells (ß-, α-IFN);

- in the *chemical synthesis of* ß- and "-interferon;
- when *introducing into a eukaryotic yeast cell a vector on the basis of which a recombinant DNA molecule with* ß and ".interferon *was designed;*
- *using monoclonal antibodies* to the corresponding IFN.

According to the nature of action and clinical significance, IFN drugs are divided into four main groups;

- *etiotropic,* acting on the causative agent of the disease;
- *immunomodulating,* correcting disorders of the immunity system that arise and develop in the course of the disease;
- *pathogenetic,* aimed at combating intoxication, dehydration, vascular lesions, organ disorders, allergic reactions, as well as the prevention of bacterial complications;
- *symptomatic,* treating associated symptoms of the disease (headache, insomnia, cough, etc.).

Antivirus tools are divided into four groups:

- chemo drugs;
- interferons;
- IFN inducers;

-immunomodulators.

Chemopreventive agents are mainly aetiotropic agents. Interferons (INF) and their inducers have a combined effect (etiotropic and immunomodulatory). Immunomodulators are used for immunotherapy and immunocorrection.

Benefits of interferon use:

- A broad spectrum of action, which is very important in the treatment of viral pneumonic enteritis of young animals;
- the effect of interferon is due to the activation of the body's natural defence mechanisms in the early stages of the infectious process;
- the effect appears immediately after the drug is administered.

The use of interferon has a *number of disadvantages-*

- short retention period in the body (up to 12 h);
- the drug is mainly used only parenterally;
- 2-3 days after the use of interferon, a decrease in the protective functions of the immune system is observed, which may lead to rapid multiplication of opportunistic microflora in the body;
- allergic reactions are possible with prolonged use.

6.8. Biotechnology for the production of interleukins

Interleukins (ILs) are substances of protein or glycoprotein nature synthesised mainly by immunocompetent cells.

Currently, 15 interleukins have been isolated and characterised.

Interleukins that play an essential role in co-operative interactions between immune-competent cells during the development of immune response include IL1, IL2, IL4, IL5, IL6.

IL 1 is a cytokine secreted by macrophages that stimulates the functions of T- and B-lymphocytes. IL 1 activates T-helper cells synthesising *IL 2,* as well as B-lymphocytes transforming into plasma cells - aitibody-producing cells. It mediates the interaction of various defence mechanisms at the level of the organism.

IL 3, a cytokine synthesised by T-lymphocytes, stimulates B-lymphocytes in humoral immune response and maturation of T-effectors in cellular immune response. IL 3 is a key factor in the development of the immune response.

IL 4 is a stimulation factor for B-cells and part of T-lymphocytes. It regulates allergic reactions of the organism. With the help of IL 4, T-helper cells influence the production of immunoglobulin E (Ig E) in the process of immune response development. The main IL 4 producers are normal T-lymphocytes and T-cell hybridomas.

IL 5 is a multifunctional cytokine that activates T- and B-lymphocytes and eosinophils; a product of activated T-lymphocytes.

IL_ is one of the earliest cntoknii produced by myohymite cells (hybridoma/plasmocyte growth factor, T-lymphocyte activating factor); it is involved in the regulation of lymphocyte functions and has an antiviral effect.

Interleukins are used as *interleukin producers*:

- cultures of normal lymphocytes or macrophages;
- . clones of transformed (tumour) cells;
- T-cell hybridomas are hybridisation products of tumour lymphocytes capable of normal growth and T-cells synthesising a specific IL;
- recombinant microbial cells obtained by genetic engineering methods.

Producers are cultured in *vUro* in vessels of a certain volume, then ILs are isolated from the culture medium, concentrated, and purified.

To increase IL production, cell cultures are stimulated with mitogens (substances that induce mitotic cell division). Globular plant proteins (phytohaemagglutinin, etc.) as well as bacterial cell wall component muramyl dipeptide are used as mitogens.

To create recombinant IL-producing microorganisms:

- genes controlling their synthesis are inserted into the microbial cell as part of the vector;
- they're multiplying a clone of recombinants;
- . cultivate the resulting progenitor in order to accumulate ILs.

E. coli cells (IL1, IL 2), *yeast-saccharomycetes* (IL. 2) are used as recombinant organisms, with yeast being more convenient, as *E. coli* cells accumulate IL intracellularly, while yeast release it into the environment, as a result of which the resulting product is more easily separated from cells and purified.

Chapter 7

PRODUCTION OF AMINO ACIDS

In industry, amino acids are produced by the following methods:

. by hydrolysis of natural protein-containing raw materials;

. by chemical synthesis;

. by microbiological synthesis,

- biotransformation of amino acid precursors using microorganisms or enzymes isolated from them (chemo-micro-biological method).

For hydrolysis can be used residues of meat processing industry (waste products of animal raw materials processing, blood, etc.), egg white, milk casein, wheat gluten, soya meal, etc. During hydrolysis protein-containing raw materials are heated with solutions of acids and alkalis to a temperature of 100...105 ^{O}C for 20-48 hours. In this case amino acids are transferred to the hydrolysate and complex, multi-stage purification is required to isolate individual amino acids. In addition, the raw material itself is considered scarce and expensive, so amino acids have a high production cost. In addition, some amino acids such as tryptophan, cysteine, methionine, tyrosine may be destroyed and racemisation may occur.

Chemical synthesis of amino acids is quite efficient, but its disadvantage is that during the synthesis process a biologically active *L-form* and a *D-isomer of* an amino acid are formed. *The* D-form is a ballast, as it is not digested by animals and humans, and some *D-forms* of amino acids have toxic properties. Separation of isomers is an expensive and time-consuming procedure. The essential amino acid methionine is synthetically produced (Fig. 9).

Fig. 9. Principal biotechnological scheme of production of microbial synthesis products

Microbiological synthesis of amino acids is the most promising and economically favourable; 60 - high-precision preparations of amino acids are obtained by this method. Its advantage is the possibility to obtain *L-amino acids* on the basis of renewable raw materials.

In recent years, biotransformation of chemically synthesised amino acid precursors using microbial cells or immobilised enzymes has been widely used.

Among the amino acid producers, yeasts (30%), ac- tinomycetes (30%), bacteria (20%) are used.

Brevibactenum flavum and *Corynebactenum glutamicum* convert more than one third of the sugars to lysine.

Microorganisms belonging to the genera *Micrococcus, Brevibactenum,*

Corynebactenum, Arthrobacter are used for selection of producers.

Glutamic acid is the first amino acid produced by microbial synthesis. It belongs to substitutable acids, has certain organoleptic properties and is widely used. Its producers are bacteria *Connebactenum glutamicum, Brevibacterium flavurnum* .

Lysine is produced by many microorganisms: bacteria, actinomycetes, blue-green algae, some species of microscopic tribes. Bacteria of the genera *Connebactenum* (C. *glutamicum), Micrococcus, Brevibactenum* are used as lnzine producers in our country.

Tryptophan is formed by microorganisms of bacterial and fungal origin: genera *Micrococcus* sp., *Candida utilis, Bacillus subtilis.*

The main consumers of amino acids are agriculture and food industry. Amino acids, most often lysine, are used as enrichments of feeds and food products of plant origin to increase their nutritional value and to balance the food on essential amino acids. In the production of mixed fodder 1 tonne of lysine saves from 40 to 50 tonnes of feed grain.

Some amino acids are used as seasonings because they have certain flavouring properties and can impart a pleasant aroma and flavour to foods. For example, glutamic cns-lota and its sodium salt (monosodium glutamate) are effective flavour enhancers for meat and vegetable dishes. This amino acid is added to many foods during canning, freezing and long-term storage.

Cystine, lysine, histidine are used to improve organoleptic characteristics of meat products, giving them a specific pleasant flavour and aroma. Cysteine and cystine with monosodium glutamate create an imitation of the smell and taste of meat, which is used in the preparation of seasonings.

Many amino acids (lysine, alanine, proline, valine and others) can relieve unpleasant food odours.

Amino acids have an original flavour and participate in the formation of flavour characteristics of food products. For example, asparagic and glutamic acids, being acidic, in neutral solutions have a very pleasant original flavour, glycine has a characteristic flavour of "refreshing" sweetness, which is intense in intensity.

Of particular interest is the sweetener aspartame, whose molecule is formed by two amino acids, phenylalanine and aspartic acid. These amino acids are synthesised microbiologically, and aspartame is synthesised from these monomers by enzymes. The sweetness of aspartame is 200 times that of sucrose.

7.1. Biotechnology of amino acid synthesis and purification

In recent years, various amino acids have been widely used in agriculture and medicine. They are of particular importance for balancing protein nutrition. Some food and feed products do not contain the necessary amounts of essential amino acids, in particular lysine. Such products include wheat, maize, oats, rice and a number of others. To eliminate the possible imbalance, amino acids are used in pure form or introduced into the composition of combined feeds produced by industry. Thus, the main area of application of amino acids should be considered the creation of diets that allow to reduce the content of vegetable proteins in feeds. It is shown that artificial mixtures of amino acids allow saving the consumption of natural forages. In addition to additives to agricultural animal feed, amino acids are used in the food industry. They are also used in the manufacture of a number of polymeric materials, such as synthetic leather, some special fibres, and films for food packaging. A number of amino acids or their derivatives have pesticidal action Methionine and γ-aminobutyric acid are widely used as medicines. The proportion of amino acids used in different sectors of the economy can be illustrated by the example of Japan, where the food industry accounts for 65% of all amino acids produced in the country, livestock - 18%, for medical purposes - 15% and other needs - 2%.

The global production of amnnoxlot currently reaches several million tonnes per year. The largest amounts of i-glutamic acid, *L-lysine*, *DL-methionine*, L-aspartic acid, and glycine are produced in the world. The main methods of obtaining amino acids are as follows: extraction from protein hydrolysates of plant raw materials, chemical synthesis, microbiological synthesis by growing cells, using immobilised microbial cells or enzymes isolated from microorganisms.

Microbiological synthesis is currently a very promising and cost-effective way to produce many amino acids. In the process of cultivation of amino acid producers, *L-amino acids* are directly synthesised. One of the important tasks of microbiological synthesis of amino acids is to obtain highly active producer strains, in particular, using genetic engineering methods. A highly active *L-threonine-producing* strain was obtained by this method.

In addition to microbiological synthesis, amino acids can be obtained by hydrolysis of natural protein-containing animal and plant raw materials. This is the most efficient method. Its main disadvantage is the irrational use of raw materials, which can be used with great benefit as protein feeds and slip products. For example, in the countries of South-East Asia monoglutamate sodium is obtained from soya meal (defatted soya flour). In the USA, a method of obtaining amino acids from wheat gluten and corn gluten, which remains after starch washing, is described. Other proteins can also be named, but their use for obtaining amino acids is economically unfavourable.

Chemical synthesis of amino acids is quite efficient, allows to obtain compounds of any structure and to organise continuous production with high automation. It mainly uses non-food raw materials and achieves high product concentration. However, as a rule, this process is multistage and requires complex apparatus. The main disadvantage of chemical synthesis is obtaining racemic form of amino acids. Sufficiently efficient and cheap ways of separating compounds into optical isomers have not yet been developed. Chemical synthesis is used to obtain only tehamniokis-lots, which can be used as racemic products. The chemistry of *ID-methionine*, the main consumer of which is the poultry industry; *L-* and *D-isomers of* methionine are assimilated by organisms equally well.

In recent years, advances in the asymmetric synthesis of amino acids have made it possible to avoid optical separation of racemic amino acids. A new approach to this problem was made possible by the discovery of homogeneous catalytic hydrogenation of olefins using rhodium complexes with phosphine ligands and the development of ways to synthesise chiral phosphines. The use of rhodium complexes with chiral phosphines as homogeneous catalysts for the hydrogenation of N-acylaminoacrylic acids allowed the asymmetric synthesis of α-amino acids with a high degree of stereospecificity and good yields.

The use of amino acids obtained by chemical synthesis for food, feed and medical purposes causes one more essential technological problem - complete liberation of the finished product from possible toxic intermediates of synthesis.

Since 2000, the combined chemical-mn-biological method of synthesis, in which the initial compound is obtained as a result of chemical reactions, and the initial stage is carried out due to the activity of enzyme systems of the corresponding strains of microorganisms, has taken a strong position.

The microbiological method of amino acid synthesis is based on the ability of many microorganisms to accumulate significant amounts of such products in the medium. Many bacteria, a number of yeasts, and other fungi have been found among microorganisms that are potential producers of glutamic acid. Most of the examined microbial strains, irrespective of their systematic position, accumulate mainly α-alanni and glutamic acid. Significantly fewer strains and in smaller amounts excrete asparagic acid, leucine, valine, isoleucine, and lysine. There is no strict correlation between the species of microorganisms and their ability to accumulate amino acids.

Despite the wide distribution of microorganisms accumulating amino acids in the growth process, there are not many producers providing economically favourable yields of these products. They are usually obtained by applying various mutagenic factors. The producer should accumulate predominantly one amino acid. The simultaneous presence of several amino acids, especially if they are close in their physical and chemical properties, makes their isolation and purification difficult.

Auxotrophic mutants of microorganisms, deprived of a number of enzyme systems as a result of mutagens, are recognised as the most valuable producers. Blockade in such mutants of the corresponding reactions in the metabolic chain leads to the supersynthesis of one of the me- tabolnts.

Common amino acid producers are gram-positive sporeless bacteria belonging to the genera *Corynebactenum, Micrococcus, Arthrobacter,* Brevibacterium and some others, but it is difficult to determine the exact taxonomic position of most of them, since the information contained in publications is obviously insufficient for this purpose.

One of the most important scientific positions of microbiological synthesis of amino acids is the question of their origin: are the amino acids in the medium the products of enzymatic breakdown of proteins as a result of autolytic process or the result of synthesis from other compounds? When using synthetic media for the cultivation of producers, it has been shown quite definitely that the amino acids found in the medium are de novo synthesis products.

Enzymatic reactions of amino acid synthesis take place inside cells. Initially, amino acids accumulate inside cells in the form of so-called free amino acids. In the early stages of culture growth, free amino acids are incorporated into the constructive metabolism of the microorganism. Active accumulation of amino acids in the medium of a periodic culture usually occurs from the middle of the exponential phase of its growth, reaching a maximum towards the end.

7.2. Amino acid production using immobilised cells and enzymes

Since 2000, the attention of researchers has been drawn to methods of amino acid production using immobilised enzymes. The method has a number of advantages: in particular, the final product is highly concentrated and pure, there is no danger of contamination by foreign microorganisms during the reaction, only natural isomers are formed as a result of synthesis, and

continuous technological processes can be carried out.
Microorganisms. -are the main sources of enzymes converted to immobilised form. Although immobilised enzymes are always more expensive than soluble enzymes, they are economically justified if even one of the following conditions is satisfied: increased stability of the enzyme, ensuring its repeated use and thereby reducing the cost of the preparation; improved product quality due to the absence of traces of enzyme in the product; prevention of undesirable side reactions.
From the already listed methods of amino acid production, one can refer to the conversion of DL-α-amino-e-caprolactam to lysine. Enzymes involved in hydrolysis and racemisation reactions can be immobilised on ion-exchange polysaccharides by covalent binding. One of the major technological difficulties in carrying out enzymatic reactions on carriers is the possibility of regeneration of cofactors (if the reaction involves them).
The enzymatic method for the production of asparagic acid from fumaric acid and ammonium is widespread due to the active stnaspartase catalysing this reaction.
Since aspartase encapsulated in a half-nacrylamide gel loses its initial activity relatively quickly, microbial cells possessing aspartase activity are immobilised. Some strains of *Escherichia coli*, whose cells are fixed in a polyacrylamide gel, have been found to be good producers of aspartase. Under optimal conditions, the yield of aspartic acid by this method reaches 12-16 thousand μM/h - g of dry cells. A significant increase in the activity of aspartase *of E. coli* cells after their immobilisation was observed.
Enzymatic methods are also used in the synthesis of L-alanine by decarboxylation of *L-aspartic* acid using *Pseudomonas dacunhae.*

7.3. Production of optical isomers of amino acids by microbial acylases

One way to separate racemate amino acids into *L-* and *D-isomers is* through an enzymatic pathway using microorganisms with specific *L-acylase* activity. Acylases break the peptide bond of acyl derivatives of amino acids or peptides, resulting in the formation of the corresponding free amino acids and peptides, as well as an organic acid.
When cultivating microorganisms containing acylases, a number of conditions should be observed. In particular, it is necessary to introduce into the media substances that are close in nature to the substrate of acylase action, for example, in the case of *DL-lysine* conversion - *ε-acetyl-L-lysine.* Acylases of fungi are located inside the cells, so their cells must be disintegrated beforehand, e.g. by ultrasound. The use of microbial acylases

for the hydrolysis of acyl derivatives of amino acids is based on the specificity of their action with respect to the optical configuration and structure of the substrate. Depending on the nature of asymmetric deacetylation of the substrate, *L-* and *D-acylases* are distinguished, with *L-acylases* being the most common in microorganisms. According to the nature of the acyl radical to be cleaved off, acetyl-, benzoyl-, chloroacetyl-, succinyl- and other amino acid derivatives are distinguished. Since acetyl- and beisoylacylases are the most common, it is these amino acid derivatives that are preferably used in reactions. The specificity of the acylase to an amino acid residue should be kept in mind. These features of acylases allow them to be used for the isolation of the desired amino acid from mixtures of amino-hyacyl derivatives.

The remaining acyl racemate and *acyl-D-amino* acid are easily separated from the optically active amino acid due to differences in physicochemical properties, such as solubility in organic solvents and water, and the ability to sorb on ions. The use of *D-isomers is* theoretically possible if racemases or chemical methods are used. For example, some N-acetyl-D-amino acids can be readily converted to N-acetyl-I-amino acids by heating in the presence of acetic anhydride. The racemisation process can also be accomplished by the use of immobilised enzymes.

Chapter 8

PRODUCTION OF IMMUNE ANTISERUM

The immunogenicity of an antigen is the ability of the antigen to induce the formation of antibodies in the body of an immunised animal.

The immunogenicity of substances is due to a number of factors:

1) molecular weight: the higher the molecular weight, the higher the immunogenicity. *Cross-linking of biopolymers between themselves and other proteins increases immunogenicity:*

- as the molecular weight of the antigen increases, its residence time in the organism increases;
- The ability of high molecular weight antigens to interact with macrophages increases significantly;

. in antigen increases both the total number of antigenic determinants and their diversity, which increases the efficiency of antigen interaction both with B- and T-lymphocytes;

2) density, location and number of antigenic determinants on the surface of antigens: as these parameters increase, immunogenicity initially increases and then begins to decrease (for example, of conjugates containing 3, 16 and 28 groups per molecule of bovine albumin, the conjugate containing 16 hapten molecules had maximum antigenicity). Only B-lymphocytes participate in the immune response against antigens that have repeating antigenic determinants; such antigens are called *independent* antigens;

3) foreignness of the immunogen: the more the antigen differs in its structure from the homologous antigen of the immunised animal, the higher its immunogenicity (*genetic foreignness of the antigen*). The less foreign the antigen, the greater the number of animals to be immunised. A mixture of antiserum against a given antigen from different animals of the same group is called a *pool.* Among laboratory animals, rabbits, guinea pigs or mice are most often used for immunisation, depending on the amount of antigen available, availability of the animal, etc. It is more convenient to immunise males, as their immunogenic response is less affected by hormonal cycles. Hens, in which antibodies are accumulated in eggs, have proved to be effective in producing antibodies against viruses. Large volumes of antiserum are obtained by immunisation of large animals: goats, rams, donkeys, horses;

4) antigen homogeneity (foreign antigen impurities may be more immunogenic than the main antigen), as this property is of great importance for the production of specific antisera;

5) The extent of the immune response depends on the amount of antigen administered. At certain concentrations of antigen, both high and low, there is

an inhibition of the humoral immune response, called tolerance, which is the reason for the choice of the *dose.* The dose of immunogen for a single injection in a rabbit or guinea pig averages 100-300 μg per 2 kgm.

The dose for sheep is 0.25-5 mg of immunogen per injection, for donkey 0.5-10.mg;

6) the method and.frequency.of.administration.of.antigen, affecting the immunological activity of antiserum (immune response is formed in the organism gradually - primary response and secondary response). Usually highly active antisera are obtained after several immunisation cycles. However, very long immunisations can lead to a decrease in specificity due to an increase in the titre of anti-telk impurity mantigeiame

Adjuvants are compounds that, when administered to the body, cause a non-specific enhancement of the immune response and thereby increase the ability of the organism to respond to any immunogen. Adjuvant properties include *oils, liposomes, bacterial cells, polymers* and others. Adjuvants introduced into the organism together with an immunogen fulfil two functions. Firstly, they promote a slower release of immunogen from the injection sites, resulting in an increased probability of immunogen encounter by symmuno-competent cells, as well as a sharp decrease in its toxicity. Secondly, adjuvants cause severe inflammation at the site of immunogen injection, activating phagocytosis and stimulating the local circulation of lymphocytes, as well as non-specific stimulation

immunocompetent cells (to enhance non-specific immune stimulation, the adjuvants additionally include a bacterial preparation.

History of adjuvants. In 1916, Le Monge published evidence that mineral oil emulsions could enhance the immune response to antigen (AG)

The term "immunological AD" was first used in the 1920s by G. Raymond to describe a substance which, when administered in combination with specific AG, produced a more pronounced immune response than pure AG.

The adjuvant activity of aluminium compounds was first demonstrated in 1926 with diphtheria anatoxin adsorbed on aluminium hydroxide, which allowed a significant reduction in antigen doses during immunisation and thus increased safety. Since then, aluminium hydroxide or aluminium phosphate has been used extensively in the development of antiserums and vaccines for human immunisation.

Freund's complete adjuvant preparation is now widely used for immunisation purposes and includes:

- a mixture of mineral oils;

. emulsifier;

-killed mycobacteria

The adjuvant preparation can be prepared in the laboratory (by mixing).

. trnparticles and mineral oil;

. one part anhydrous lanolin;

- four-partO,15MK phosphate buffer;
- mycobacterium preparation to a final concentration of 10 mg/ml so that the mycobacterium particles are evenly distributed throughout the volume.

AD is mixed with aqueous immunogen solution in the ratio of 2 :1 to form a non-dispersible emulsion in which the aqueous immunogen solution is in micelles.

The use of adjuvant in immunisation;

- reduces the possibility of tolerance; _
- allows the range of administered immunogen to be extended from 50 to 200 MCG per injection.

Granulomas often form after adjuvant administration. In case of deterioration, the next injection should be skipped and only after the animal has recovered should immunisation be continued.

8.1. Methods of immunisation

It is most effective to inject immunogen in small portions into a large number of points.

Intradermal injection. Several scratches are made with a scalpel on the skin of the animal cleaned of hair and then the soluble immunogen is rubbed into this area.

The advantage of this method is that a high immune response is observed after a single injection, resulting in a significant reduction in the consumption of immunogen, and the disadvantage - in the labour-intensive execution and strong painful sensations in the animal from extensive areas of ulceration in places of immunogen injection.

Subcutaneous immunisation. 5-6 portions of approximately 2 ml of immunogen solution are injected into points along the animal's vertebral column

Intramuscular injection. Simultaneously inject part of the immunogen into the muscle of the hind legs of the animal in small portions.

Intraperitoneal administration. This method is used for the immunisation of small laboratory animals such as mice or sea lions.

Direct injection of immunogen into the lymph nodes. Immunogen is injected into the lymph nodes located in the hamstring fossa of the rabbit's hind legs. As a result, the amount of immunogen is reduced to 10-50 μg, and the volume of injected solution - to 25 μl. The method is technically complicated (skin incision, search for lymph node and immunogen injection without damaging it).

Intravenous injection. Usually used for repeated injections followed by blood sampling of the animal. The immunogen solution is injected directly into the bloodstream. Adjuvant is not used in this method of antigen administration as the animal may die.

Injection of immunogen into the pads or conjunctiva of the eye. Immunisation is started with immunogen mixed with complete Freund's adjuvant and partial Freund's adjuvant is used for repeated injections. Prior to blood sampling, 1-3 intravenous injections are given 7-9 days in advance to raise antibody levels. Causes severe pain in the animal.

During immunisation, small blood samples are taken from the animals to assess the amount of antibodies.

The maximum level of immune response to the administration of a large number of soluble antigens is reached 40-60 days after the first injection.

When the immunogen is microbial cells, the highest possible antibody levels are observed earlier.

After the end of the first immunisation cycle, the animal is allowed to recover for 30 days and re-immunised (1-3 intravenous injections).

8.2. Production of guinea pig antiserum to porcine insulin

Multipoint injections of insulin subcutaneously, intramuscularly and intraperitoneally are performed according to the following scheme.

1)1, 8, 15 days - 200 mcg of insulin in 0.5 ml of physiological solution mixed with 0.5 ml of complete Freund's adjuvant and administered to the animal, followed by 30 days of rest;

2) Day 45, 46, 47 - 150 mcg of insulin in 0.7 ml of physiological solution;)

3) Blood sampling is done after 7-9 days from the heart.

The second cycle of immunisation is carried out after 30 days of rest of animals according to the schemes45, 46 and 47 days.

To obtain antiserums with titres satisfactory for enzyme-linked immunosorbent assay (ELISA), 4-, 5-fold repeat cycles of immunisation are often required.

Blood sampling is carried out each time on days 7-9 after the last injection.

Immunisation of rabbits with thyroxine conjugate with bovine serum albumin (BSA). Multi-point injections are carried out along the spine and intramuscularly in the area of micro-nodes of the hind legs according to the following scheme:

D.l- day- 1-2 mg of T4-BSA conjugate inO.7ml of physiological solution is mixed with 0.7ml of Freund's adjuvant and the resulting emulsion is administered to the animal;

4) 31, 32, 33rd or 45, 46, 47th days - intravenously administer 0.7-1 ml of physiological solution containing.1-2 mg of T4-BSA.

Blood is drawn from the heart on days 7-9 after the last injection.

After one month, the intravenous immunisation cycle is repeated.

Blood collection and antiserum production. The immunised animal is used as an immune serum donor for 5-7 months (5-6 immunisation cycles).

Animals that have undergone several cycles of immunisation are called hypo-immunised.

Blood is withdrawn from the ear vein, from the heart (by cardiac puncture) in the volume of 50-70 ml in a rabbit and 5-10 ml in a guinea pig into sterile tubes rinsed with sterile buffer solution.

Blood devoid of cellular elements is called *plasma.* It contains fibrinogen, resulting in the formation of a *fibrin clot* throughout the tube, which is gently removed by centrifugation at 1000-2000 rpm for 15 min. Plasma devoid of fibrin is called *serum.*

To remove the proteins of the complement system, the serum is heated for 30 min at 56 °C' at which time the antibodies remain active.

Processing of blood and obtaining serum should be done with the utmost care, avoiding destruction of erythrocytes.

The presence of intracellular proteins and enzymes in serum may result in additional background in some modifications of the immunoassay.

8.3. Storage of antiserum

The native immune serum can be stored for 3-6 months without loss of immunological activity in a frozen state at -20 °C, pre-filled into 0.5-1 ml vials. Freezing/freezing leads to a decrease in the immunological activity of the serum.

The antiserum is conveniently stored lyophilised in ampoules under vacuum. In dry form it retains its immunological activity at room temperature for 1-2 years.

Thawed or diluted dry serum in solution can be stored at +4 ^{O}C for a week, having previously added a preservative - chloroform, 0.1% sodium azide or 0.01% mertnolate.

Preservatives can be inhibitors of marker enzymes in im- munoenzyme assays.

Aluminium compounds are currently the most widely used *adjuvants* in vaccines. In addition to these, other substances such as emulsions, liposomes and nanosomes are also used as adjuvants in vaccine production

Emulsions are "oil-in-water" and "*water-in-oil*", such as Freund's incomplete

AD.

Liposomes are synthetic spherical particles consisting of a bimolecular layer of lipids that hold the antigen in the capsule and perform a dual action, acting as a carrier of AG vaccine and immunomodulator.

Virosomes- liposomes that include besides the *viroid* (viral capsid) more proteins that promote the fusion of liposome and cell membranes.

Novel immunomodulatory molecules include:

- CpG-containing oligonucleotides (CpG)l;

. monophosphoryl lipidA(MPA);

. soap tree bark extract (QS21).

CpGs (*CpG-oligodeoxynucleotides*), short single-stranded synthetic DNA molecules containing cytosine C, guanine G, and the P-phosphodiester component of DNA, act as danger signals for cells of the vertebrate immune system and activate the innate and acquired immune response. CpG containing oligodeoxynucleotides are significantly more active compared to adjuvant-Freunda.

QS-21 is a purified plant extract that increases the ability of the immune system to respond to antigens. It is extracted from the bark of the soap tree (*Quillaja saponaria*). The extract contains water-soluble triterpene glucoside compounds related to xaxaponenes.

The adjuvant system AS04 consists of the immunomodulator MI.-monophosphoryl lipid A, an *immune system stimulant adsorbed on a-w compounds.*

The vaccine containing the adjuvant AS04, authorised for human use, is a hepatitis B virus surface AG vaccine. It produces a high titre of specific antibodies and increased seroprotectin levels with fewer doses than the classical hepatitis B vaccine in combination with aluminium compounds.

With the AS04 adjuvanted vaccine, protective antibody levels are maintained longer with fewer non-circumvented doses.

The vaccine against uterine neck cancer, human papillomavirus (HPV) types 16 and 18, also contains the AS04 adjuvant system.

8.4. Antibody isolation and purification

The γ-globulin or immunoglobulin (Ig G) fractions of the immune serum are usually used to increase the relative amount of antibodies. The most complete isolation of the γ-globulin fraction without reducing its immunological activity is achieved by precipitation with ammonium sulphate, after which the precipitate is dialysed for a long time with frequent changes of buffer solutions.

Another widespread method is precipitation with polyethylene glycol with MM = 4000-6000. In 10% polyethylene glycol solution all proteins with MM = 150 000 aggregate, resulting in precipitation of proteins of γ-globulin fraction, while proteins of lower molecular weight remain in solution. This method allows the preparation of the γ-globulin fraction of the antiserum to be obtained very quickly. The method is quite effective if the preparation is not stored at low temperatures. Prolonged storage of the preparation with preservation of immunological activity is possible after careful dialysis.

The antibodies in the majority of antigens belong to the KlgG fraction of mune serum. In order to increase the sensitivity and specificity of the assay, in many cases it is useful to use the native serum IgG fraction for sorption on solid phase and for the preparation of conjugates. The simplest and most readily available method for the isolation of IgG is ion-exchange chromatography on DEAE-Sephadex or DEAE-cellulose.

Ig G-fraction isolation of rabbit antiserum:

1) 10 ml of rabbit antiserum was diluted 2-fold with 0.15M NaCl, 6.26 g Na_2SO_4 was added, mixed and incubated for 12-16 h at (4

2) The precipitate is removed by centrifugation, dissolved in 10 mL phosphate buffer and then dialysed against the same buffer for 24 h at room temperature;

3) After removal of the precipitate by trifugation, the solution is transferred to a column with DEAE-cellulose equilibrated with phosphate buffer;

4) is determined by measuring the optical density of the elu

ata at 280 nm, the concentration is calculated using the molar absorption coefficient e = 1.5 g/L cm*.

To isolate rabbit Ig G, 50 mM phosphate buffer containing 20 mM ethylenediamine tetraacetate #8 g Na_2HPO_4 - H_2O and 7.4 g EDTA-Na dissolved in 900 ml of water, bring the pH to IM NaOH and then add water to 1 litre.

Detection of antibodies by immunosorption. The specificity of the serum is tested in reactions with the antigen preparation used for immunisation and a selection of chemical compounds close to it in structure. In the case of insufficiently purified antigen, non-specific reactions are possible due to the presence of antibodies to the antigens used. Cross reactions with other compounds can be observed in the presence of chemical structures close to the structure of antigenic determinants of the immunogen.

The removal of non-specific antibodies from antiserum is usually accomplished by adsorption on the appropriate non-specific antigen. Adsorption of non-specific antibodies in the precipitation reaction with a soluble fraction by centrifugation may result in the appearance in the antiserum of either soluble adjuvant-antibody complexes (AD-AT) or an excess of added AD, the presence of which affects the ability of the antiserum to react with the specific antigen in ELISA. It is most convenient to use antigen immobilised on a solid phase for adsorption of antiserum. By adding such an immunosorbent to the antiserum or by passing the antiserum through an immunosorbent column, cross-reacting antibodies can be quickly freed. The method of affinity chromatography on immunosorbents is used to obtain preparations of purified antibodies. Immunosorbents based on CNBr-activated Sepharose are the most widely spread. CNBr-activated agarose produced in our country is not inferior to its foreign analogue in its main parameters.

The main operations used to obtain antigens or antibodies immobilised on CNBr-agarose are given in Table 8.

Table 8

Covalent binding to CNBr-activated Sepharose

Operation	Terms and conditions
1. Weigh out the required amount of CNBr-activated Sepharose	1 g of dried preparation gives about 4.5 ml of gel
2. Wash on a glass filter and allow to swell to a gel	Take a solution of 1 mM HC1 for washing and then borate or hydrogen carbonate buffer for equilibration and swelling
Operation	Terms and conditions
3. Dissolve the protein in the buffer	Use hydrogen carbonate (0.1 M $NaHCO_3$ containing 0.5 M NaCl) or reverse buffer
4. Mix the protein solution with the gel suspension	The solution should contain 5-10 mg of antigen
5. Block the remaining active groups	Two hours at room temperature at +4 °C for 12 h the gel was washed with buffer containing blocking compounds, e.g. 1 M ethanolamine or 0.2 M glycine, pH 8.0
6. Wash away unbound antigen	The buffer in which covalent binding was carried out is used, followed by 0.1 M acetate buffer, pH 4, containing 0.5 M NaCl, after which the gel is again equilibrated with borate or hydrocarbonate buffer

It should be noted that when working with immunosorbents based on CNBr-activated Sepharose, buffer containing amino groups should not be used

Obtaining antibody fragments. When developing some modifications of ELISA, it may be necessary to use enzyme conjugates not with the whole immunoglobulin molecule, but with its fragment, which specifically binds to the antibody molecule - Fab-fragment. This is primarily due to the fact that the Fc-fragment of the molecule, responsible for the effector functions of immunoglobulin, has the ability to interact nonspecifically with other proteins sorbed on carriers during solid-phase ELISA. The efficiency of specific sorption is determined by the nature of the antigen, the concentration of protein sorbed on the carrier, and a number of other factors.

Obtaining Fab fragments

1) a solution containing 0.1-3 mg of P_2-fragment in 0.45 ml of 0.1 M 0.1 M Ya-phosphate buffer, pH 6.0, is prepared;
2) 0.05 ml of 0.1 M 2-mercaptoethanol in the same buffer containing.5 mM EDTA is added;
3) the mixture is incubated at 37 ^{O}C for 1.5 h;
4) antiserum tests.

Antiserums obtained even from the same animal vary considerably in their ability to bind antigen. For comparative characterisation and quality assessment of antiserums, they are tested, which allows two important tasks to be solved: firstly, to select only those sera that meet the requirements of immunoassay by their properties; secondly, to standardise antiserums during their industrial production in long immunoassay kits.

The primary selection of antiserum is based on finding its titre, which is an integral parameter characterising the interaction with the antigen. More detailed information about the serum is obtained by determining the affinity and concentration of antibodies. The next step is to determine the antigenic specificity of the antibodies, i.e. their ability to interact with structurally similar antigens.

The fundamental difference between these steps is that in titre determination the binding of a selected concentration of antigen at different dilutions of antiserum is analysed. When determining the affinity and concentration of antibodies, the binding of the antiserum to different concentrations of labelled antigen is analysed. When studying the specificity of the antiserum, different concentrations of cross-reacting antigen are added at the appropriate dilution and at a constant concentration of labelled antigen.

In all cases, antigen concentrations are used either close to the minimum detectable concentrations in the assay or in the range that corresponds to their concentration in the sample.

Determination of antiserum titre. Titre is an effective value characterising

the binding capacity of antibodies, depending on their concentration and affinity. The absolute value of the titre also depends on the method and conditions of the experiment and on the initial concentration of free antigen.
Quantitatively, the titre is found as the limiting dilution of serum at which the positive effect of interaction of antiserum with antigen registered by this method is still observed (for example, if the titre is established by the method of precipitation of free antigen in a gel, and for the first serum the formation of a precipitate is observed at a dilution of 2 times, and for the second - 4 times). Sometimes the term "50% titre" is used, implying a dilution of serum that causes 50% binding of antigen. The working titre is the initial serum dilution used directly in the experiment (Table 9).

Schematic of serum testing

Table 9

Stage of research	Purpose
Titre	Selection of highly active immune antisera, determination of final and working titres
Affinity	Assessment of affinity and concentration of high-affinity antibody fractions
Specificity	Determination of antibody specificity

The titre of the antiserum depends strongly on the concentration of the antigen used for testing and on the method of determination. For example, the same serum may have a titre of 100 in an immunoprecipitation test and a titre of 200,000 in an enzyme-linked immunosorbent assay. Therefore, when testing serum and determining the titre, it is best to use the same method as the assay and the antigen concentration should be close to the "minimum in the range chosen for the assay. In addition, it should be borne in mind that both homogeneous and heterogeneous methods of determining antigen concentration are used in immunoassays, which may differ greatly in the structure of the immune complexes formed. In particular, in solid-phase methods, the probability of formation of cyclic antigen-antibody complexes is much lower than in homogeneous methods, and consequently, the observed affinity of antibodies in both systems is much lower.
When developing ELISA methods, the value of the antiserum titre determined in the indirect method is usually used, in which antigen is initially sorbed on the walls of microplate wells and then the binding of immune serum to it is studied in serial dilutions. The binding of the antibody under investigation to the immobilised antigen is recorded using an antispecies conjugate to the immunoglobulin fraction of the serum of the animal used for immunisation.

The results in this case are usually evaluated using the concept of "50% titre". The maximum signal recorded from the optical density of the product of the reaction of oxidation by hydrogen peroxide of 5-amnos-licylic acid is -1.4 optical units. The serum titre is the dilution at which the optical density recorded by ELISA has a value close to 0.7.

This approach is usually used at the first stage of quality assessment of the obtained immune sera and allows selection of highly active immune sera. However, for a more in-depth evaluation of the obtained antibodies, their affinity is determined When developing ELISA methods for a number of antigens present in biological fluids in low concentrations of the order of, antiserums should used that have not only a high titre, but also a sufficient content of antibodies with a high binding constant.

Antibody Affinity. Immobilised antibodies and radioactively labelled antigens are mainly used in such experiments.

Titration of antiserum to insulin by indirect solid-phase ELISA: on the abscissa axis is the dilution of serum diluted to SOO times, on the ordinate axis is the optical density at 430 nm of the product of peroxidase oxidation of 5-aminosalicylic acid by hydrogen peroxide. Blood serum of a non-immunised animal was used as a control.

When enzyme tags are used, the chemical structure of the labelled antigen may differ significantly from that of the unlabelled antigen. In these cases, the binding constants obtained generally characterise interactions in this particular system, so that an immune serum that has a high affinity in reaction with a conjugate of one structure may have a significantly lower binding constant with a conjugate obtained by a different method.

If a researcher is dealing with the development of competitive ELISA methods, i.e. has a set of immune sera and an antigen-enzyme conjugate, an important step in the development of a test system is the selection of an antibody-conjugate pair characterised by the required interaction constant. For screening, the γ-glo-bulin fraction of each antiserum is immobilised on the surface of a microplate and its binding to the available set of conjugates is studied. The effective binding value is the inverse of the conjugate concentration at which 50% of the active centres on the carrier are bound.

Obtaining such a conjugate for a wide range of antigens is not an easy task. Enzyme-labelled antibodies are a more universal reagent, so let us focus on a general approach to determining the affinity of antibodies in serum or ascites based on solid-phase ELISA. This method, proposed by B. Frite, allows us to determine the binding constant of antibodies to antigen during their interaction in solution.

Method formulation involves two axioview stages.

1) performing the antigen-antibody reaction in solution;

2) Determination of antibody concentration after equilibration by solid-phase ELISA.

Chapter 9

BIOTECHNOLOGY VACCINE PRODUCTION

The development of immune biotechnology has been driven by the need to obtain a large number of biologics for the prevention, diagnosis and treatment of both infectious and invasive diseases. Using the achievements of immuology, microbiology, biotechnology and genetic engineering, the modern veterinary and medical bioindustry obtains and produces a wide range of biopreparations, including vaccines, diagnostics, microbial antagonists (probiotics), antibiotics, immune sera and their derivatives. The share of animal vaccination in the system of anti-epizootological measures is determined by many factors, including "the availability of effective and harmless biopreparations. Veterinary statistics show that there is a tendency to increase the use of specific means of animal defence for many infectious diseases. For this reason, the development of the modern bio-industry of immunopreparations is caused by the practical need to obtain a large number of immunopreparations (in terms of volume and list) for prophylaxis, diagnostics and treatment of animals.

The current immunoprophylaxis arsenal includes a number of vaccines that differ:

- by type and nature of production technology, method of application and efficacy, namely: *live, inactivated, corpuscular, chemical anatoxins;*

. number of antigens: *monovaccines and associated.*

The majority of these vaccines *are* still produced by proven classical methods and only some of them are designed according to new, unconventional principles - *these are artificial antigen vaccines, subcellular (ribosomal), subunit and genetically engineered vaccines.*

Vaccine production is divided into several technological steps, including the *accumulation of biomass* of organisms or their products, *concentration and purification of antigens,* in some cases their *inactivation, addition of sorbent and adjuvant substances, lyophilisation, packaging,* etc.

In all cases, a universal *scheme of* principle for the production, testing and introduction of new vaccines is used. It consists of the following stages:

. The author has obtained a vaccine (attenuated) or production (virulent) strain;

. Author's preparation of the vaccine in laboratory conditions and experimental study of sterility, toxicity, harmlessness, reactogenicity and immunogenicity *in vitro* and on animals;

. experimental and production production of the experimental series of the preparation, assessment of harmlessness of antigenic and immunogenic

properties in laboratory conditions and in limited epizootological and epizootological conditions of production;
- State commission testing of a new biopreparation in order to determine its harmlessness, reactogenicity, immunogenicity and epizootological efficacy in a wide production experience. Approval of normative documents (technical conditions) for the manufacture and control of vaccines and instructions for their use.

9.1 Live vaccines

Live vaccines are immunopreparations containing hereditary modified forms of infectious agents (vaccine strains) that have lost virulent properties but retained the ability to induce immunity.

As a rule, live vaccines are prepared from attenuated (elavirulent) strains of pathogenic microbes, which are not capable of causing disease but retain the property of multiplying in the organism of the inoculated animal and thus causing a vaccine reaction - a benign vaccine process in the form of local specific inflammation or general asymptomatic infection. Such a reaction in both cases of local or general exposure should cause a specific restructuring of the immunoreactivity of the organism and the formation of immunity.

Vaccine strains proposed for the production of live vaccines.must:

- *be avirulent* or weakly avirulent strains, irrespective of the method of production. These properties must be sufficiently robust in the production and use of the vaccine. Therefore, the live vaccine must be harmless, which in turn depends on the virulence and genetic stability of the vaccine strain, as well as on the general immunological status of the vaccinated animals;
- have *high antigenicity and immunogenicity.* For live vaccines, the criterion of immunogenicity is the ability to induce intense immunity in at least 70% of once vaccinated animals;
- have the *ability to multiply in specific organs* as well as short-term persistence in the inoculated animal;
- are carriers of *genetic stability of phenotypic properties,* especially low virulence and high antigenicity, even in rapid successive passages in naturally susceptible animal species;

- possess *genetic markers* to distinguish them from epizootic (field) strains of pathogens (more than two and independent of each other);

. be characterised by the *absence of infectivity* (contagiousness) in case of excretion of the vaccine strain from the body of the vaccinated animal, and be harmless to other animal species;

- have *stability during storage* and a wide range in dosing between the minimum and maximum immunogenic doses. The vaccination (optimal) dose of a vaccine is the number of live microorganisms contained in a unit volume. The volume of the vaccination dose should be neither too large nor too small to ensure convenient administration to the animal and accurate dosing.

Vaccine virus strains must have certain *titres of activity* (infectivity) in a specific biological system (macroorganisms, bird embryos, cell cultures, etc.). To determine the immunogenicity and the degree of immunity in vaccinated animals use carefully studied with stable properties control (virulent) strains of the relevant pathogens against which live and inactivated vaccines are used. Vaccine strains that fulfil the above requirements are no longer pathogens. The preservation of such strains and the maintenance of their properties are the responsibility of the relevant government agencies.

Attenuated microorganisms with high residual virulence and infectivity that do not fulfil the above criteria for vaccine strains are "pseudovaccine" and for them "vaccine affinity" is very conditional.

Principles of attenuation of bacteria and viruses. The term "*attenuation*" (Latin *attenuatio* - reduction) means artificial persistent weakening, reduction of virulence of pathogens of infectious diseases.

The phenomenon of attenuation was discovered in 1880 by L. Pasteur. The attenuation of viruses and bacteria is widely used in the preparation of live vaccines, or, to be more precise, in the selection of strains intended for the production of live vaccines. Attenuation of microorganisms is carried out by various methods based on the adaptation of the pathogen to the organism of non or low susceptible in natural conditions of animals or on the adaptation of the microorganism ".unfavourable conditions of cultivation in which it is exposed to chemical, physical, biological and other factors. Any method of attenuation reduces the virulence of the microorganism for naturally susceptible animals, and this newly acquired property of the pathogen must be hereditary.

Examples of vaccine strains that are attenuated by exposing microorganisms to physical or chemical factors include:

- The virulence of L. Pasteur and L. S. Cenkovskii weakened by growing cultures of the pathogen at high temperature (42.5 °C);
- capsule-free strains of sybireaemia vaccines (STI) (N. NGins-burg) or GNKI (S. G. Kolesov), obtained by directed exposure to cultivation conditions and subsequent strain selection;
- STI vaccine strain, obtained by inoculation of virus culture on dishes, is a capsule-free avirulent immunogenic mutant STI-1;
- A strain for the HNCI vaccine produced by alternating crosses on clotted

normal horse serum and peptone-free agar, resulting in a weakly virulent capsule-free mutant, Shchuya-15;

- French scientists A. Calmette and C. Guerenoslabnlibn the virulence of the tubercle bacillus by long-term (230 passages over 18 years) cultivation on potatoes with bile, resulting in an avirulent BCG vaccine strain;
- *Br. abortus* vaccine strain was obtained by American scientists D. Beck and W. E. Cotton through long-term cultivation of the virulent strain in potatoes.

Vaccine strains of ^viruses are obtained, as a rule, by passaging them in the body of the same animal species that is not the natural host of the virus. In this case, the virulence of the virus for the other animal species is increased, while for the host it is weakened and becomes sufficiently constant and low, which makes it possible to use this strain as a vaccine strain. Thus, L. Pasteur obtained an anti-rabies vaccine by repeatedly passing the street rabies virus through the brain of a rabbit, as a result of which the virulence of the virus for the rabbit increased sharply and its virulence for other animal species, as well as for humans, was lost. The English scientists S. Iyer and N. Lobson obtained an attenuated strain of avian pseudoplague virus (Newcastle virus - strain H) from which they prepared a live vaccine against this disease by successive passages on chicken embryos. G. Hudson achieved a significant weakening of virulent swine fever virus by passages through rabbits; this weakly virulent strain is used as a live virus vaccine. American researchers H. Kaprowski and J. Bey have similarly attenuated swine fever virus. Kaprowski and J. Baker.

The method of passaging microorganisms through species that are not natural hosts of the pathogen is also applicable to bacteria. For example, D. F. Konev weakened the culture of the swine rye pathogen by passaging it through the organism of rabbits and then testing its virulence on pigeons. The Kazan Veterinary Institute proposed as a vaccine strain *Br. melitensis* No. 89, which was attenuated by 25-fold passage through the organism of a chicken embryo.

Attenuation can be induced by exposure to bacteriophage, antibiotics, radiant energy. Attenuated strains can be obtained by selection of dissociating cultures. Natural attenuation of infectious agents is possible in nature and in laboratory conditions during storage of a collection of microorganisms. Detailed data sheets are prepared for all strains of microorganisms applicable for the manufacture and control of bioproducts (vaccines, anatoxins, diatostikums). The data sheets contain: virulence limits, beyond which these strains are unsuitable for the manufacture of these preparations; types of agricultural, commercial and laboratory animals susceptible to them in natural conditions; the main ways and methods of infection of these animals; information on the duration of incubation period, as well as on the degree of danger to humans; a detailed description of the serological properties of strains, which indicates the time of formation of specific antibodies in vaccinated animals and their titers

Live vaccines are considered to be among the most effective vaccine preparations, due to the high survival rate of vaccine strains and the more complete preservation of a set of protective antigens in them. Immunisation with live vaccines is not difficult and in most cases is a single shot. However, live vaccines can cause postvaccine complications and are difficult to standardise and store.

9.2. Inactivated corpuscular vaccines

Inactivated corpuscular vaccines, in contrast to live *vaccines*, are produced from pre-selected or freshly isolated virulent and immunogenic strains.

The State Veterinary Medicines Control Service has certain requirements for strains of microorganisms that can be used for the manufacture of specific immunoprophylaxis products, including inactivated corpuscular vaccines. Such strains are called *production* strains. Along with them, bioindustry institutions have the same strains of microorganisms used only to infect experimental animals for controlling the activity of vaccines and immune sera. These strains are called *control* strains. Production and control strains of microorganisms must be classified, i.e. their species affiliation must be established. their species affiliation must be established, cloning must be established, they must represent a homogeneous population originating from a single cell or virus-containing plaque, with genetically fixed features characteristic of it: morphological (size, shape, pigmentation of the colony; size and shape of cells or virions; presence or absence of flagella, capsules, spores); biochemical (acquisition or loss of the ability of cells to use certain amino acids, growth factor, ability to utilise carbohydrates); antigenic (loss of certain antigens or their presence); other features (e.g. degree of sensitivity to physical and chemical factors affecting the bacterial cell or virion).

The appearance of dissociants in the progeny of even a single microbial cell or virion is considered to be a general biological law. Therefore, dissociated forms of microorganisms are allowed in the population of production and control strains of microorganisms in an amount not exceeding 5% of the total number of cells (virions), provided that they do not adversely affect the phenotypic properties of the strain.

Strains intended for the production of inactivated vaccines, anatoxins or diagiostikums must represent a culture of a particular genus and species of microorganisms and must have a clearly defined set of properties that allow the strain to be assigned to that genus and species. In vivo, such strains must retain high virulence (a measure of pathogenicity). Virulence is determined in each case on specific susceptible large and laboratory animals, on developing avian embryos or cell cultures. Virulence is expressed as lethal dose (LD) 50, infectious dose (ID) 50, tissue cytopathogenic dose (TOD) 50.

The production strains used for the preparation of inactivated vaccines, anatoxins and diagosticums should be identical in antigenicity and other properties to the majority of cultures circulating in natural conditions that cause infectious diseases (epizootic strains). In addition, such strains of microorganisms should have high immunogenic properties, i.e. cause a state of immunity to infection with live virulent cultures of the species or serological variant of the corresponding farm and laboratory animals. The most reliable information on the immunogenic properties of the strain is obtained by direct infection of laboratory animals or naturally susceptible farm and domestic animals previously vaccinated with this strain. These requirements should be common to all production and control strains, regardless of the preparation based on them. In addition to these basic requirements, there are additional requirements for strains used in the manufacture of live and inactivated vaccines.

Depending on the methods of inactivation of production strains of microorganisms, a distinction is made:

. warmed wpkupny obtained by heating daily broth culture of microbes at a temperature of 56...60 ^{O}C for 0.5-2.0 h;

- *anavaccines* obtained by exposing the culture to formalin (0.3-0.5 %) and heat (39...40 °C) for 28-30 days.

In addition, depending on the chemical used to inactivate the cells, there are *alcohol, acetone, $_{CHL}$orophor$_{M}$oene, $_{K}$arbo$_{(L)oene}$,* and *$_{K}$arbo$_{L}$oene* vaccines

To obtain inactivated (killed) vaccines, it is necessary to preserve the antigenic properties of the original culture, which requires complex nutrient media and gentle methods of microbial inactivation.

Killed corpuscular vaccines are less effective than live vaccines and must be administered multiple times to achieve robust immunity. This prolongs immunisation for a long time, which is particularly undesirable in unfavourable epizootic conditions.

Inactivated vaccines have a large amount of ballast substances that cause allergic and other adverse reactions and abnormalities during administration.

9.3. Chemical vaccines

Chemical vaccines are prepared from molecular antigens extracted from the microbial cell by thelymically (more often by chemical means or ultrasound), and are therefore "referred to as molecular vaccines. They include vaccines derived from soluble derivatives of the microbial cell, such as toxins, but are described as cannatoxins.

Chemical (molecular) vaccines can be produced either by biosynthesis or chemical synthesis. So far, only the biosynthetic method has been used in practice since the 1940s. However, experiments in recent years show that chemically synthesised vaccines can be used quite soon. Undoubtedly, modern chemical vaccines, obtained from the synthesised antigens, are still inferior to live vaccines in terms of immunogenicity, but at the same time, their use reduces the danger of postvaccinal complications, and they are less reactive, more standard, it is possible to concentrate antigens in small volumes, adsorb them on various substances, use adjuvants and thus increase the antigenic properties of biopreparations.

Moreover, *chemical vaccines* can be used in associated preparations, i.e. it is possible to design multicomponent preparations against many pathogens.

In order to create chemical vaccines, it is necessary first of all to decipher the structure of the microbial cell, especially the specificity of anti-hepatic structures. Immunological antigens are immunologically active substances of different chemical nature, capable of generating specific immunity when introduced into the organism. Protective antigens consist of those structural elements that ensure pathogenicity of the microorganism, but at the same time determine antigenicity and specific immunogenicity. They are located either on the surface of the microbial cell, on the cytoplasmic membrane (CPM) or in the cell wall, and may also be extracellular products of microbial cells.

Chemically, protective antigens are macromolecules. They are high molecular weight polymers - glycoproteins, protein-palisaccharide-lipid complexes. All microorganisms - bacteria, viruses, rickettsia - share the same nature of protein antigens. It is known that antigenicity of a substance is largely determined by its macromolecularity. It, in turn, depends on the molecular weight of the antigen. When designing chemical vaccines, there is a possibility of enlargement of antigen molecules by aggregation and formation of covalent chemical bonds. It is generally accepted that polymerisation of antigen is one of the practical ways to increase its immunogenicity.

Obtaining a chemical vaccine by biosynthetic means represents:

- selection of an appropriate microorganism that maximises the production of protective antigen;
- selection of a nutrient medium that ensures biomass growth and release of the protective antigen. The most suitable are synthetic and semi-synthetic media (no ballast);
- development of cultivation conditions, including deep cultivation, in high-speed bioreactors;

. development of methods of industrial extraction of antigen from the biomass and its purification, allowing to preserve the original antigenic structure of the preparation and to ensure a sufficient degree of purity of the preparation with a satisfactory content of antigen (extraction of biomass of whole "lysed cells: enzymatic lysis, prolonged freezing and rapid thawing, sounding, mechanical processing, alkaline lysis).

Purification of antigen is carried out depending on its properties and isoelectric precipitation with acids and alkalis, salting with neutral salts, alcohol precipitation, sorption and elution, ultrafiltration, chromatographic methods, etc. are used.

Antigens are isolated from animal and inactivated cultures. The same physical and chemical methods are used for inactivation as for killed vaccines (formaldehyde, phenol, glutaraldehyde, hydrochloric acid hydroxylamine, hydrogen peroxide, aznd sodium hydroxide).

The main principle of antigen production is the use of gentle physicochemical methods, possible reduction in the number of steps in the technological chain and increase in the immunogenicity of the preparation. It is of fundamental importance to ensure harmlessness of the vaccine, its sterility (filtration, 0.3% formalin at 40 ^{O}C for 5 days, ß-propiolactone, etc.). For antigen sorption, aluminium hydroxide gel is more often used, which also has an adjuvant effect. Sorbed vaccines create a depot in the body, from which antigens are gradually absorbed, ensuring the pro-located effect of the preparation. This makes it possible to reduce the frequency of vaccination while maintaining a high prophylactic effect.

9.4. Anatoxins and associated vaccines

Chemical vaccines also include *anatoxins, which are* preparations derived from bacterial exotoxins. They are completely devoid of toxic properties, but retain antigenic and immunogenic properties.

The anatoxin *method was developed* and proposed by the French scientist A. Ramon in 1923. The first stage of anatoxin production is the cultivation of toxigenic cultures of bacteria causing toxaemic infections in affected animals (tetanus, botulism, anaerobic cancer infections, staphylococcus, etc.).

Microorganisms are cultured in liquid nutrient media under a specific regime for maximum accumulation of exotoxin, then the exotoxin is separated from the bacteria and neutralised (inactivated). To obtain anatoxin (according to Ramon), the toxin is combined with formalin (0.3-0.5%) at 39...40 ^{O}C for 28-30 days. The efficacy and harmlessness of the anatoxin depend on the degree of purification. The presence in the preparation of bacterial proteins, proteins of the nutrient medium increases its reactivity and sensitising properties. Purified anatoxins are largely devoid of these negative properties. In addition, they can be concentrated in a small volume of liquid and adsorbed. Currently, ß-propio-lactone is widely used as a detoxifying agent with formaldehyde. Treatment of the toxin with proteolytic enzymes such as pepsin, trypsin, prolase and papain followed by filtration on Sephadex Y-100 and Y-200 columns has also been proposed. The efficacy of anatoxins depends not only on the quality of the antigen, but also on the form of the preparation and the method of administration.

Anatoxins can be prepared as *liquid, dry, emulsified or sorbed preparations.*

Association vaccines are preparations containing several synergistic, balanced antigens. Their advantage over monovaccines is the possibility to ensure simultaneous production of immunity against several infections by administering a single complex preparation. The condition for con- struction and production of associated vaccines is epizootological indicators for simultaneous immunisation against the corresponding diseases, the possibility of eliminating the competitive effect of antigens in association, data on the absence of increased reactogenicity and sensitising activity of a complex vaccine compared to a mono-drug. The use of associated vaccines facilitates the organisation of immunoprophylaxis, reduces the number of injections and in some cases increases their immunogenicity and epizootological effectiveness.

9.5. New principles of vaccine design

Since the 1970s, new trends in the field of vaccine design have emerged. Along with the improvement of classical technologies of traditional (old) vaccines manufacturing, the creation of artificial vaccines, or vaccines of artificial antigens, was intensively developed. At the same time, two fundamentally new types of vaccines were produced - from individual structures of microorganisms, or subcellular vaccines from bacteria, and subunit viral vaccines, as well as genetically engineered vaccines of artificial antigens.

The development of artificial antigen vaccines involves the synthesis of analogues of natural antigenic determinants responsible for the induction of immunity. The immunogenic properties of artificial antigenic determinants are determined by two factors: their location on the surface of the molecule and their conformational structure.

Practically, the process of obtaining artificial antigens is folded up:

- from the isolation of the biologically active antigen;

. deciphering the homomolecular structure;

• artificial resynthesis by chemical or genetically engineered means.

This method is promising for the development of vaccines against pathogens whose antigens have immunogenic and protective properties.

There is *öea type of modern artificial vaccines :*

• of artificial immunising molecules synthesised on the principle of natural analogues:

- conjugated antigens (hapten carrier);
- synthetic proteins;
- synthetic polyamino acid antigens;
- Conjugents of synthetic polyamino acid antigens with synthetic adjuvants;
- Synthetic;
- of artificial immunising molecules that have no analogues in nature: artificial complexes or conjugates of antigenic determinants with a synthetic non-natural molecule with immunomodulatory action.

The fundamental novelty of the modern stage of vaccine design is not only in the fact that the creation of artificial macromolecules possessing the necessary antigenic determinants has been initiated.

Immune response genes (Ir genes) have now been discovered and their action has been shown to occur at the level of. T-system lymphocytes. The main thing is to find ways to phenotypically correct the control of immunogenesis. As a result, an immune response can be obtained even if the organism does not react to a given antigen. For this reason, the most promising way in the field of creating new vaccines is recognised as the design of such micromolecular complexes, which include both the necessary antigenic determinants and the carrier structure providing phenotypic correction. In other words, the *immunogenicity of vaccines can be increased by using specialised adjuvants.*

Currently, synthetic low molecular weight compounds - chemical analogues of bacterial structures with immunoregulatory activity - have been created.

The first of this type of compound to be synthesised was *muramshdi-peptide* (TIR). Murein, a complex macromolecular compound with a strong immunomodulatory effect, is known to form the basis of the bacterial cell wall. TIR causes a reduction in the humoral response to vaccine administration. This suggests that the most promising vaccines are those presented with an antigen linked to a synthetic immunomodulator.

As a carrier of antigenic determinant having the properties of adjuvant, such non-natural polymeric associations as, for example, poltzvcnylpcrtsds (a class of synthetic electrolytes) can be used. They enhance the function of T-helper cells and promote the interaction of B- and T-lymphocytes, increasing the number of antibody-braising cells.

To date, the effects of other *adjuvants,* such as *Freund's adjuvant, lipopolysaccharide (LPS*), which, like TIR, stimulate T-lymphocyte activity; *BCG and LPS*, which activate macrophages, have also been deciphered.

The first fully synthetic vaccines against diphtheria toxin, streptococci, influenza virus, hepatitis B virus were created on the basis of conjugation of antigenic determinants (bacterial and viral) with synthetic carriers (TIR). Synthetic antigens of FMD virus and subcellular (ribosomal) vaccines were produced. In recent years, the development of a new type of vaccine preparations consisting of ribosomes of the corresponding pathogen was started. These vaccines are called ribosomal or subcellular vaccines. They have some advantages over live vaccines or isolated microbial cell antigens.

The immunogenicity of bacterial ribosomes was discovered in 1965. Investigating the protective activity of mycobacterial subcellular fractions, E. Doumans and other scientists isolated ribosomes containing rRNA and experimentally proved their immunogenicity. These results served as an impetus for further research and development of ribosomal vaccines in a number of countries (vaccines against salmonellosis, escherichia, streptococcus, haemophillus, francisella, brucella, leptospira, etc.). The following techniques are used to disintegrate the bacterial cell:liquid press, mechanical destruction, ultrasound treatment. Ribosomes are isolated and purified by different modes of differential ultracentrifugation or ammonium sulphide precipitation. Treatment of ribosomes with detergents, polyethylene glycol is also used.

The advantages of ribosomal vaccines include the following features:

- bacterial ribosomes are not toxic to animals and are low reactive;
- vaccines have a more pronounced immunogenicity compared to scorpuscular vaccines;

. ribosomal vaccines are able to induce cross immunity to different serotypes and serogroups within the pathogen species.

The mechanism of action of ribosomal vaccines has not been definitively deciphered. According to researchers, in mycobacterial ribosomal vaccines, the protective effect is exerted by bacterial RNA, which is reproduced in cells by reverse transcriptase (similarity to RNA viruses).

Thus, a new hypothesis of bacterial resistance independent of antibodies and cellular immunity factors is proposed. However, it is not possible to prove such an effect in other cases. On the contrary, in other experimental infections, the protective effect of ribosomal preparations is clearly immunological in nature, involving both humoral and cellular immunity factors. The administration of highly purified preparations of RNA bacteria also proved ineffective.

The second hypothesis about the mechanism of action of ribosomal vaccines was formulated on the basis of data obtained during the study of the effect of the vaccine on S. *typhimuriwn,* and according to it, the main immunogenic factor is the species-specific ribosomal protein 30S.

There is a third opinion: the composition of ribosomal preparations often contains antibodies of the bacterial cell envelope - capsular polysaccharides, cell wall proteins, etc. One can assume immunogenic effect of these "contaminating" agents as well.

Thus, the question of the protective factor of ribosomal vaccines is very controversial and far from being solved. It is not clear whether ribosomes fulfil the role of a carrier adsorbing the antigen molecule and changing the conditions of its contact with immunocompetent cells, or whether they are adjuvants. The formation and joint action of RNA-antigen complexes seem to be the most probable, which requires special unknown conditions. Immunological specificity is determined by the antigen, and RNA participates in the presentation of the antigen to immunocompetent cells or influences cellular immunity.

At present, ribosomal vaccnaprotnv of human *pneumonia pathogens - Str.pneumoniae, H. influenzae, KLpneumoniae, Str.piogenes* - has been developed. Klebsiella peptidoglycan was used as adjuvant. A vaccine from Shigella zonne was prepared.

9.6. Subunit viral vaccines

Live and inactivated vaccines in many cases provide reliable protection of the animal organism. However, they are not without disadvantages - postvaccine complications, allergic reactions, etc. may occur in animals. That is why work was started to improve viral vaccines.

Subunit vaccines are a further way to improve split-virus vaccines.

In order to obtain them, the antigens required for immunisation must be available in purified form. For this reason, the action of subunit vaccines is directed against those infectious agents for which the antigenic structure is known, the protective antigens have been identified, and they can be separated and concentrated as subunits of the biopreparation. In the process of forming immunity to viruses, external proteins are the most complete antigens. It is this fact that has been used as the basis for the development of subunit vaccines.

The most promising model for creating subunit vaccines is the so-called enveloped viruses, which have a lipoprotein membrane. Subunit vaccines can be classified as the third generation of inactivated viral vaccines. The traditional vaccines are culture vaccines (first generation). Modern vaccines are characterised by a high degree of purification and concentration (third generation). A typical example of the use of different types of vaccines is human influenza vaccination. Today, several types of inactivated purified vaccines are used - sub-virion vaccines made of completely destroyed virions (split-virus vaccines) and sub-virion subunit vaccines with the highest degree of purification $_{CI}$.

Thus, subunit vaccines against animal influenza and parainfluenza should contain haemagglutinin and neiramidase as subunits. Subunits of FMD, adeno- and herpesvirus pathogens have been identified and isolated in pure form, against which subunit vaccines have also been produced. However, it should be remembered that viral subunit vaccines are active only when they induce virus-neutralising antibodies.

The technological process of manufacturing subunit vaccines consists of the following steps:

. obtaining virus from infected cell cultures or embryos;

- Purification of the virus using a plate separator or on a Sephadex column;

.microfiltration;

. concentration (about 50 times) in filters;

- additional purification by ultracentrifugation in a gradient heatthiostnssaccharose;

. cleavage of concentrates by detergents;

- Extraction of virus glycoproteins by ultracentrifugation or ultrafiltration;
- dialysis; . ...
- Sterilising filtration of the subunit concentrate;
- Standardisation of the subunit concentrate;

. biopreparation control.

Bacterially subunit vaccines are also being developed, in which purified H- and O-antigens from the cell wall of the capsule, CPM, flagella, pili, etc. are used as protective subunits. Vaccines against pneumococcosis, pasteurellosis, colibacillosis and a number of other diseases have also been obtained. All followers note the biotechnological difficulties in obtaining subunit vaccines and, at the same time, the promising scientific direction and the great potential of this type of biopreparation.

9.7. Genetically engineered vaccines

In the 70s of the XX century, the successes of genetic cell engineering allowed to develop a new technology for the production of pro™- vovirus vaccines, called genetically engineered vaccines. The need for such developments was dictated by the lack of natural sources of raw materials (suitable animals) and the impossibility to reproduce the virus in classical objects (tissue culture, etc.).

The principle of genetically engineered vaccines includes:

- isolation of natural antigen genes or their active fragments;
- the insertion of these genes into simple biological objects - bacteria, yeast;
- obtaining the required product in the process of cultivation of the biological object producing antigen.

The genomes of viruses are negligible in size compared to the genome of a cell (prokaryotic or eukaryotic). The genes encoding protective proteins can be cloned directly in uDNA-containing viruses, or in rRNA-containing viruses after reverse transcription of their genome (for viruses with a continuous genome) or even individual genes (for viruses with a fragmented genome).

At the first stage of development of new biotechnology, scientists were mainly engaged in cloning viral genes encoding the synthesis of proteins carrying the main antigenic determinants. Soon recombinant bacterial plasmids carrying genes or genomes of hepatitis B, influenza and poliomyelitis viruses were obtained.

The next step was to obtain the antigen. The issue proved to be complicated, as the expression of viral genes in the prokaryotic system was not very important. This can be explained by the fact that viruses have adapted during evolution to parasitise in the human body. However, over time, antigen expression has been obtained. One of the most typical examples showing the necessity of creating gene-in-

The problem is that no cell or animal cultures sensitive to the virus have yet been found, so the development of a genetically engineered method of vaccine production has become a necessity.

The method consists of cloning the genome in *E. coli* cells using plasmid and phage vectors. Bacteria carrying recombinant plasmids produce proteins specifically reactive with antibodies against the virus itself. In 1982, the first

experimental vaccine against hepatitis B was produced in the USA. Eukaryotic cells (yeast, animal cells) are also used to produce virus-specific proteins (antigens).

Intensive work is underway to create other genetically engineered vaccines, in particular against influenza, herpes, foot-and-mouth disease, tick-borne encephalitis and other viral infections. A novel approach to viral vaccines is the incorporation of genes responsible for the synthesis of viral proteins into the genome of another virus. In this way, recombinant viruses are created that provide combined immunity. In conclusion, it should be emphasised that all vaccines, regardless of their origin, must meet several basic requirements, without which they cannot be used for active immunisation of animals, namely: possess specific sterility, harmlessness, standardity, moderate reactogenicity, and have high antigenic and immunogenic activity.

9.8. Vaccine control

Vaccines of all types are tested after preparation according to the following parameters:

1) *sterility* (inactivated) or purity of growth (live) is controlled by sowing on nutrient media;

2) harmlessness is tested by administering the vaccine to laboratory animals. The vaccine must not cause illness or death of animals;

3) The vaccine is administered to a group of laboratory animals and after a period of time sufficient for the development of active immunity (15-20 days), this group, together with a control group of unvaccinated animals, is infected with a known lethal dose of the pathogen. The control animals should die, while 80% or more of the vaccinated animals should survive.

In some cases, the immunogenicity of the preparation is judged by indirect indicators: the number of agglutinins in vaccinated animals (leptospirosis vaccine), antitoxins in PH (botulism vaccine);

(I) *Reactogenicity* - adverse local and general reactions to vaccine administration that are evaluated in animals and in human vaccination;

5) *pyrogenicity is the* ability of a substance to cause an increase in body temperature;

6) *control of* dry vaccine against swine rye vaccine of VGNKI from strain BP-2 is carried out as follows. For purity control, the dry (lyophilised) vaccine is diluted with sterile phy- enological solution in the ratio of 1: 10. Swabs are prepared from the bacterial suspension, Gram stained and microscoped. In the preparation should be typical small Gram-positive bacilliform cells in the absence of postoroninskh mncroorganisms Simultaneously carry out VYSE TM H"a"C? ^(KiTta)_ (Tar0tstszi And agar Sab)^ (p0)- Seeds

are maintained at 37...38 ^{O} C 10 days, cultures for fungi - at 20...25 (^{O})C 15 days. The growth of extraneous microflora in the specified terms on all nutrient media should be absent in the presence of typical growth on MPA and in MPB culture of the rye pathogen. In order to control the vaccine for harmlessness and activity 20 white mice weighing 17-18 g are injected subcutaneously with 0.2 ml of the preparation. The vaccine is considered harmless if no more than 5 mice die. After 14 days, all surviving vaccinated and five control mice are infected with a culture of a virulent strain of the swine rye pathogen at a known lethal dose. The vaccine is considered active if the control mice die within three to four days and at least 75% of the vaccinated mice survive.

Causes of Biological Reaction to Vaccine :

. incomplete attenuation of the viral or bacterial component. This is rare, but is a known occurrence and should be reported to the vaccine manufacturer;

. there are animals that are particularly susceptible to infection with an attenuated pathogen. For example, a live vaccine may be administered to a small puppy, kitten or immunosuppressed individual who has recently had an immunosuppressive disease (feline leukaemia, distemper, etc.);

• the vaccine is administered in the wrong way. Live "feline influenza" vaccine attenuated for subcutaneous administration often causes infection when administered intranasally (cat licks the injection site or inhales liquid from a full syringe);

• During vaccination, the animal is infected or incubating infection; . . . . _

. the animal is a carrier of a clinically latent infection. Vaccination stress can cause release of the pathogen and the development of clinical symptoms, e.g. a cat with latent herpesvirus-1 infection;

. usually live vaccines may induce mild clinical symptoms. Subcutaneous administration of live vaccine often initiates mild systemic clinical symptoms such as depression, lethargy and lack of appetite that last about 24 hours. Intranasal vaccine often causes mild symptoms of respiratory disease, such as sneezing and sometimes slight discharge from the eyes and nose.

Chapter 10

VITAMIN PRODUCTION

10.1. General characterisation of vitamins

N. I. Lunin, a Russian physician, who was the founder of the doctrine of vitamins, established in 1880 that when white mice were fed only artificial milk consisting of casein, fat, milk sugar and salt, the animals died. Consequently, natural milk contains other substances essential for nutrition. In 1912, Polish doctor K. Funk, who proposed the name "vitamins", summarised the experimental and clinical data accumulated by that time and came to the conclusion that such diseases as scurvy, rickets, pellagra, beri-beri are diseases of nutritional deficiency or avitaminosis. Since then, the science of vitamins (vitaminology) began to develop rapidly, which is explained by the importance of vitamins not only in combating many diseases, but also in understanding the essence of a number of life phenomena. The method of vitamin detection applied by Lunin (special diet for animals - provoking experimental avitaminosis) was the basis of the study. It was found out that not all animals need a full complex of vitamins..Some of their species can independently synthesise certain vitamins. At the same time, many mould and yeast fungi and various bacteria develop on artificial nutrient media only when added to these media extracts from plant or animal tissues containing vitamins. Thus, vitamins are essential for all living organisms.

The study of vitamins is not limited to their detection in natural products by biological tests and other methods. Active preparations of vitamins are obtained from these products, their structure is studied, and finally they are synthetically produced. The chemical nature of all known vitamins has been investigated. It turned out that many of them occur in groups of 3-5 or more related compounds, differing in details of structure and degree of physiological activity. A large number of artificial analogues of vitamins were synthesised in order to elucidate the role of functional groups. This has contributed to the understanding of the action of vitamins. Thus, some derivatives of vitamins with substituted functional groups have the opposite effect on the organism compared to vitamins, competing with them for binding with specific proteins in the formation of enzymes.

Vitamins have letter designations, chemical names or names characterising them by their physiological action. In 1956, a unified classification of vitamins was adopted, which has become generally used.

The availability of chemically pure vitamins has made it possible to elucidate their role in the metabolism of the organism. Vitamins are either part of enzymes or components of enzymatic reactions. In the absence of vitamins in the organism, the activity of enzyme systems in which they participate, and therefore metabolism, is disturbed. There are several hundred known enzymes, which include vitamins, and a great number of reactions catalysed by them. Many vitamins are mainly participants in the processes of decomposition of food substances and release of the energy contained in them (vitamins B_{15} B_2, PP, etc.). They also participate in the processes of synthesis: B_6 and B_{12} - in the synthesis of amino acids and protein metabolism; B_3 (pantothenic acid) - in the synthesis of fatty acids and fat metabolism; B_c (folic acid) - in the synthesis of purine and pyrimidine bases and many physiologically important compounds - acetylcholine, gluta-thione, steroidovidr.

Thus, vitamins are of great physiological importance. The elucidation of the physiological role of vitamins made it possible to use them for vitaminisation of foodstuffs, in medical practice and in animal husbandry. Vitamins became especially widely used after the development of their industrial synthesis.

Vitamins are not formed in heterotrophs. Only autotrophic organisms have the ability to synthesise them.

Almost all known vitamins can be obtained by microbiological methods. However, it is more economically feasible to obtain vitamins by isolation from natural sources or by chemical synthesis. Microorganisms can be used to produce complex vitamins such as ß-carotene (provitamin A), B_2, B_{12} and vitaminar precursors.

Vitamin B_{12} (cyanocobalamin) in animal tissues has a very low concentration of vitamin (in bovine liver 1 mg/kg) in order to use this source for industrial purposes The chemical syi- thesis is very complex. Acetic acid bacteria, fungi and propionic acid bacteria are capable of synthesising vitamin B_{12} *The Propionibacterium* and *Pseudomonas (P. denitnficans)* are of greatest industrial importance.

Vitamin B_{12} concentrate is intended for enrichment of animal feed. For enrichment of fermented dairy products with vitamin B_{12} propionic acid bacteria are used both in pure form and in the form of concentrate prepared on milk whey.

Vitamin B_2 (riboflavin) can be extracted in small amounts from natural raw materials. It is more abundant in carrots and cod liver. The most active producers of vitamin B_2 are yeast-like fungi of the genus *Eremothecium ashbyii,* belonging to the class of ascomycetes, as well as bacteria *Bacillus subtilis*. Vitamin B_2 is enriched in some types of white bread and is used to

colour foodstuffs orange-yellow.
Carotenoids are precursors of vitamin A, of which ß-carotene is the most active. Carotenoids are not synthesised in the human body and must therefore be supplied from outside. In the liver, carotene is converted into vitamin A. Carotenoids can be produced by mushrooms and yeasts. In industry, ß-carotene is most often produced by the microscopic fungus *Blakeslea trispora.*
ß-Carotene is used in the manufacture of food products as a colouring agent It is used in the production of sausages instead of sodium nitrite to ensure high intensity and stability of colour, in the production of lollipops, food pastes, cupcakes and other confectionery products. In many countries, ß-carotene is used to colour butter. In addition, it has antioxidant properties that help to extend the shelf life of the product.
Vitamin^ is industrially synthesised by the yeast *Saccharomyces serevisiae,* it is used for the treatment and prevention of rickets ~canine animals
Microbiological production of biotin used in the diet of chickens and pigs is promising. Currently in the West, most pig feeds include biotin obtained by chemical synthesis. Chemical synthesis results in a racemic mixture, and only the D-form of the vitamin, which is synthesised by microorganisms, is biologically active.
There are 40 large industrial producers of vitamins in the world; 18 of them are in the USA, 8 in Japan, 14 in Western Europe. The Swiss concern *Hoffman La Roche,* which produces 50-70% of all vitamins, holds the leading position in the production of vitamins.

10.2. Production of water-soluble vitamins

Vitamins are mainly produced synthetically, and only in some cases, individual steps in the chain of synthesis are carried out by biological means. The production of vitamin concentrates from plant or animal products has lost much of its significance.
Obtaining vitamins refers to a fine organic multistep synthesis. The following water-soluble vitamins are synthesised by chemical methods: B,, B_2, B_3, B_6, B_c, C, PP, and B_{12} by enzymatic methods of microbiological synthesis. Fermentation is also used in one of the stages of vitamin C synthesis. This vitamin in the form of an individual crystalline substance of high purity is formed by the reduction of D-glucose into D-sorbitol. The latter is enzymatically oxidised into *L-sorbose*, which after a series of operations is converted into vitamin C (Fig. 10.1).
The production of **vitamin B** (thiamine) (Fig. 10, *6)* is based on the condensation of 2-methyl-4-amino-5-chloro (bromo) methylpyrimidine with

4-methyl-5-3-oxyethylthiazole. The coenzyme of vitamin B1, cocar-boxylase (Fig. 10, 7), or thiamine diphosphoric ester, used for the treatment of heart diseases, is obtained by phosphorylating thiamine with subsequent purification on ion-exchange resins and crystallisation.

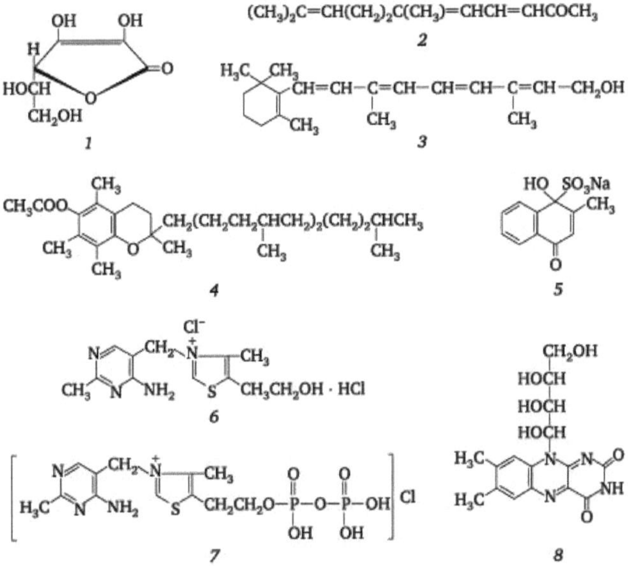

Figure 10. Structure of water-soluble vitamins:
1 - vitamin C (ascorbin); *2* - linoleic acid; *3* - vitamin A (retinol);
4 - vitamin E (tocopherol); *5* - vitamin K; 6 - vitamin B_1 (thiamine);
7 - thiamine phosphoric ester; *8* - riboflavin (B_2)

Vitamin B_2 (riboflavin) (Fig. 10, *8*) is formed by cultivation of *Eremothecium ashbyii* and other microorganisms without isolation in the form of dry biomass (used only for feeding farm animals), and synthetic riboflavin (used in medicine) is obtained as a crystalline product by destructive oxidation of *D-glucose* (from corn starch) into *D-arabonic* acid and by a number of other operations transformed into the final product - yellow-orange crystals of high purity. An important derivative of riboflavin, its coenzyme riboflavin-5'-phospho-phosphate sodium (R= Na) (Fig. 11, *1*), used for injections, is obtained by phosphorylation of riboflavin, while another coenzyme, FAD, is obtained by phosphorylation of riboflavin.

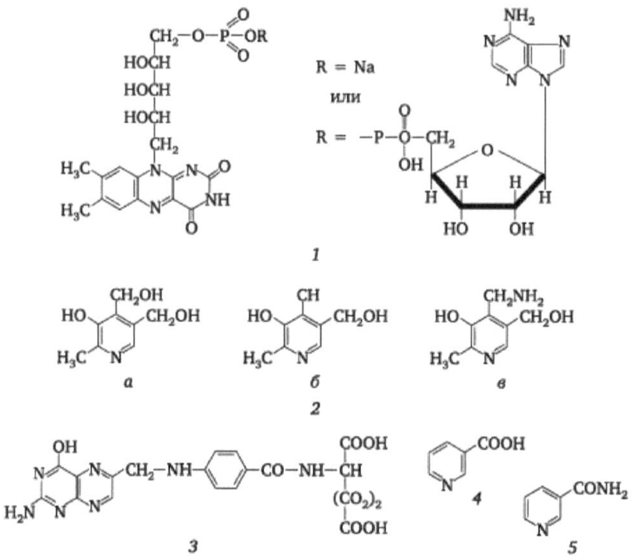

Figure 11. Structure of coenzyme forms of vitamins:
1 - riboflavin-5 -sodium phosphate, R - FAD; *2* - vitamin B_6 (*a* - pyridoxine; *b* - pyridoxal; *c* - pyridoxamine); *3* - folic acid ($B_{(c)}$);
4 - nicotinic acid (PP); *5* - nicotinamide

(R is an adenosine-5'-phosphate residue) (Fig. 11, *1*)-create by condensation of riboflavin-phosphate and adeiosin-5'-phosphate.

Vitamin B_2 is found in the cells of various microorganisms as a coenzyme in flavoproteins (primarily the corresponding enzymes from the class of oxidoreductases - FMN, FAD). Therefore, riboflavin (flavoproteins) can be produced by bacteria, yeasts and filamentous fungi. The most tempting strains are those that produce 0.5 g or more of riboflavin in 1 litre of medium on liquid media. Such organisms include *Ashbyii gossypu, Eremothecium ashbyn,* and *Candida guilliermondn.* Given the variability of active producers of these species in their ability to synthesise vitamin B_2, a systematic selection of cultures is necessary during their exploitation in production. Usually, active producers of the first two species form bright orange colonies on agarised media. Genetic engineering methods were used to obtain a strain of hay bacillus that produces about 6 g of riboflavin per 1 l of medium including molasses, protein-vitamin con- centrate-hydrolysate.

The high riboflavin yield of *E. ashbyn* correlates with purine nitrogen and other nitrogen sources, the content of which should be sufficient. Glucose or sucrose are used as carbon sources, and yeast and corn extracts, soybean meal, oil (fat) are practised. The fermentation medium usually includes corn and soya flour, sucrose, corn extract, potassium dihydrophosphate, calcium carbonate, sodium chloride and unsaturated fat.

Fermentation is usually carried out for 5 days at pH 5.5-7.7. Once sucrose has been used (after about 30 h), vitamin B_2 begins to accumulate markedly, first in the mycelium and then in the culture liquid. The whole biomass can be dried, ^{AND}-G product with a residual moisture content of 8 %, containing 1.5-2.5 % riboflavin, 20 % protein, thiamine, nicotinic acid, pyridoxine, cyancobalamin, trace elements and other substances, is recommended for animal feeding.

In case of high riboflavin yields, the vitamin can be isolated in its individual state and used in medicine along with synthetic riboflavinomics.

For *Candida* $_{gut}H_{i}e_{(rmOn)}d_{(a)}$, it is important to regulate the iron content of the nutrient medium; optimal concentrations range from 0.005 to 0.05 μg/ml on average. Certain yeast strains can produce up to 0.5 g/litre or more vitamins in 5-7 days. However, more productive species and strains of fungi - *E. ashbyii and Ashbyiigossypii* - are preferred for the purpose of industrial production of riboflavin

Vitamin B_c (folic acid) (Fig. 11, *3*) is synthesised by the one-step condensation of 2,4,5-triamino-6-oxypyrimidine, 1,1,3-tri-chloroacetone and *η-aminobenzoyl-L-glutamic acid.*

Vitamin PP (nicotinic acid) (Fig. 11, $^{4)}$ is obtained by oxidising ß-picoline (extracted from coal tar), which has limited resources, and by oxidising quinoline or 2-methyl-5- ethylpyridine. In addition to nicotinic acid, nicotinamide is used for medical purposes (Fig. 11, *5).*

Vitamin B_3 (optically active *D-pantothenic* acid) is used in medicine as a calcium salt:

$$HOCH_2C(CH_{(3)})_2CH(OH)CONH(CH_2)_2COOH.$$

For the needs of animal husbandry there is no need for separation of pantolactone racemate into optical antipodes at intermediate synthesis steps. Synthesis of racemic calcium pantothenate consists in al- fractional condensation of isobutyral and formaldehyde with subsequent transformation into pantolactone, then in its condensation with ß-alanine, leading to "formation of the final product.

Vitamin B_6 (pyridoxine) (Fig. 11, *2, a*) is synthesised by condensing methoxyacetyl acetone with cyanoacetic ester in the presence of ammonia to form 2-methyl-4-methoxymethyl-5-cyano-6-oxypyridine, which is subjected to nitration, then converted to pyridoxine by a series of operations. Another method for the preparation of pyridoxine is also known, via 4-methyl-5-propoxyoxazole by diene synthesis with buten-2-diol-1,4 formal. Other forms of

B_6 are pyridoxol (Fig. 11, *2, b)* and pyridoxamine (Fig. 11,[2], *c)*

Vitamin B12 (cyancobalamin) is obtained only by microbiological synthesis. Its producers are prokaryotes, primarily propionic bacteria, which also naturally produce this vitamin. Mutants of *Propionibacterium shermami* M-82 and *Pseudomonas denitrificans* M-2436 produce up to 58-59 mg/l of cyanocobalamin on liquid medium.

Given the important function of the vitamin in the human body (it is an anti-anaemic factor), its global production has reached 10 tonnes per year, of which 6.5 tonnes are used for medical needs, while Z.5 tonnes are used in animal husbandry.

Domestic production of ciancobalamin is based on the use of *P. freudenreichii* var. *shermami* culture, cultivated in batch mode without access to oxygen. The fermentation medium usually contains glucose, corn extract, ammonium and cobalt salts, pH around 7.0, maintained by adding NH,OH; fermentation duration is 6 days; after 3 days, 5,6-di-methylbenzimidazole (precursor of vitamin $B_{12)}$ is added to the medium and fermentation is continued for another 3 days.

Cyanocobalamin accumulates in bacterial cells, therefore one- raction for the isolation of the vitamin consists of the following: cell separation, extraction with water at pH 4.5-5.0 and temperature 85 90...C. in the presence of a stabiliser (0.25 % iatriiitrite solution). C. in the presence of a stabiliser (0.25 % iatrium titrate solution). Extraction proceeds for an hour, after which the aqueous solution is cooled, neutralised with sodium hydroxide solution, protein coagulants - trivalent iron chloride and aluminium sulphate - are added, followed by filtration. The filtrate is evaporated and additionally purified using methods of ion exchange and chromatography, after which the vitamin is crystallised at 3-4 ^{O}C from a one-acetone solution.

Crystalline cyancobalamin can be obtained by using zorcin or phenol to form adducts with it, which are relatively easy to decompose into their constituent components.

When implementing this biotechnological process, one should not forget about the high photosensitivity of vitamin B_{12}, so all operations should be carried out in darkened conditions (or under red light). On acetone-butyl and alcoholic bards with the addition of cobalt salts and methanol in our country is obtained fodder preparation KMB 12 - concentrate, containing vitamin B_{12} and other rose substances.

Vitamin C (ascorbic acid) is an anti-scurvy vitamin found in all higher plants and animals. Only humans and bacteria do not synthesise ascorbic acid, but humans do not need it and bacteria do not need it. Nevertheless, certain species of acetic acid bacteria are involved in the biosynthesis of *L-sorbose*, a semi-product of this acid. Thus, the whole process of ascorbic acid production is mixed, i.e. chemo-enzyme- tiveTM (m))- .

The biological stage of the process is catalysed by membrane-bound polyoldehydrogenase. The last stage, chemical, includes the following steps: condensation of sorbose with diadetone to yield diacetone-i-sorbose, oxidation of *diacetone-L-sorbose* to diacetone-2-ketogulonic acid, which is then hydrolysed to yield 2-keto-1-gulonic acid; the latter is enolysed and further transformed into *L-ascorbic* acid (Fig. 12).

Fermentation of *Gluconobakter oxydans* is carried out on media containing sorbitol (20 %), corn or yeast extract, under intensive aeration (8-10 g O_2/l/h). The yield of *L-sorbose* can reach 98 % in one to two days. When the culture reaches log-phase, sorbitol can be additionally added to the medium, bringing its concentration to 25 %. It was also found that *G. oxydans* can also oxidise higher concentrations of polyalcohols

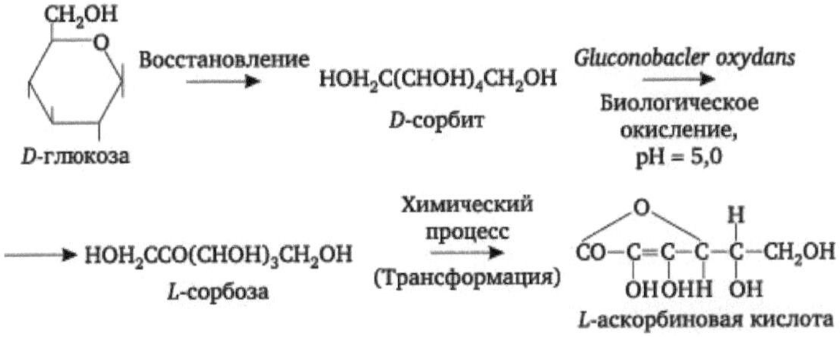

CH_2OH
D-глюкоза
Восстановление
$HOH_2C(CHOH)_4CH_2OH$
D-сорбит
Gluconobacler oxydans
Биологическое окисление, pH = 5,0
$HOH_2CCO(CHOH)_3CH_2OH$
L-сорбоза
Химический процесс
(Трансформация)
L-аскорбиновая кислота

Recovery
D-sorbitol
L-ascorbic acid
Gluconobacler oxydans
Chemical process
Biological oxidation

Fig. 12. Scheme for production of ascorbic acid (vitamin C)

Alcohols (30-50"/^ created in the last stages of the process. This is due to polyoldehydrogenase contained in the cell biomass. The possibility of obtaining *L-sorbose* from sorbitol using immobilised cells in a polyacrylamide gel (PAAG) has been proved in principle.

10.3. Production of fat-soluble vitamins

Production and use of ergosterol. In industry ergosterol is obtained using yeasts *Sacch. cerevtsrae, Sacch. carlsbergensis*, as well as mycelial fungi.

Sowing is done with a large amount of minocultures. Cultivation is carried out at high temperature and strong aeration in a medium containing a large excess of carbon sources in relation to nitrogen sources.

Yeast, *Aspergillus* and *Pemcilhum* fungi are used to produce crystalline vitamin D_2 or concentrate. Irradiated dried yeast is used as a concentrate in livestock farming.

The absorption maximum of ergosterol is noted at 280 nm. It is this radiation that excites the individual bonds of the A and B rings in the ergosterol molecule and causes its transformation into vitamin D_2. Irradiation is carried out by UV lamps with a wavelength of 280-300 nm (dry yeast) or in a thin layer of 5% yeast suspension. With shorter-wavelength and longer-wavelength radiation, the yield of other compounds of sterol nature increases. The yield of vitamin D_2 (and the formation of other compounds) is affected by the duration of irradiation, temperature, and the presence of impurities. For this reason, irradiation of ergosterol used as food additives should be done with great care.

The USSR industry produced a preparation called "Feed hydrolytic yeast enriched with vitamin D_2". In 1 g of absolutely dry yeast contains 5000 IU of vitamin D_2, at least 46% of crude protein and essential amino acids, including lysine, methy-they"-tryptophan.

To obtain crystalline vitamin D_2, yeast or mushroom mycelia are hydrolysed with hydrochloric acid solution at 110 °C. The hydrolysed mass is treated with alcohol at 75...78 °C and filtered after cooling to 10-15 °C. The filtrate is evaporated to 50 % solids and used as a concentrate of group B vitamins.

Vitamin0 2 is obtained from the mass remaining *after* filtration. The mass is washed, dried, crushed and treated twice at 78 ^{O}C with three times the volume of alcohol.

The alcohol extracts are thickened to 70 % solids content. In this way a lipid concentrate is obtained. It is saponified with NaOHa solution and sterols remain in the unsaponified fraction. Ergosterol crystals precipitate out of solution at 0 ^{O}C. The crystals are purified by recrystallisation by successive washing with 69% alcohol, alcohol and benzene mixture (80 : 20) and recrystallisation. The obtained crystals of ergosterol are dried, dissolved in ether, irradiated, after which the ether is distilled off and the vitamin solution is concentrated and crystallised.

To obtain an oil concentrate, the vitamin solution after filtration is diluted with oil to a standard level.

The source of ergosternia can be myceln of hornbeams, which remains as a waste product of the antibiotic and citric acid industries.

The industrial production of ergosterol has been established using a lipid fraction utilising *n-alkanes*. The dry mass of yeast was extracted with petroleum ether to extract residual hydrocarbons, and ergosterol, ubiquinone-9 and other fat-soluble compounds were isolated from the resulting lipid fraction (microbial fat), a by-product of the microbiological industry.

Ergosterol-enriched, UV-irradiated yeast is used in animal husbandry as a feed additive.

Ergosterol is a STARTING product for the production of some steroid hormones, medical and food preparations, the amount of which is insufficient for the needs of the national economy, and the introduction of new production facilities is the task of the near future.

Production of carotenoids. Carotenoids are produced by chemical synthesis and by isolation from natural sources - plants and microorganisms. ß-carotene, vitamin A, ß-ano-8-carotial, ethyl ester of ß-ano-8-carotenoic acid, canthoxanthin and a number of other carotenoids are chemically synthesised and synthesised on a factory scale. Some plants (carrots, pumpkin, grass, rose hips, sea buckthorn) are also traditional sources of carotenoids.

Mycelial fungi and yeasts are increasingly used for the same purposes. Bacteria and algae are also of interest as carotenoid producers.

Some photo-trophic bacteria are promising in this respect, in which carotenoid yield can be regulated depending on light intensity. In Japan, carotenoid-rich purple bacterial biomass is used to supplement the diet of chickens to enhance the colour of the yolk. Carotenoids can also be obtained in significant quantities from some algae (e.g. *Spongiococcus excentncum, Chlorella sp.).*

Among chemotrophs, *the* yeasts *Rhodotorula gracilis, R. rubra, Rhodospondium diobovatum,* as well as ac- *tinomycetes* (*Ad. chrestornycetes*™. *aurantioiäeusAdchrysornallus* var. *Carotinoides*), *mycobacteria* (*Mycobacterium phlei, M. carotenum*), fungi *(Mucoraceae, Dacrymycetaceae*, etc.) are used for *carotenoid* production. Of interest are some strains of *Flavobacterium* synthesising the pigment zeaxanthin, which cannot yet be obtained by chemical synthesis.

The heterothallic mucor fungi *Blakeslea tnspora* and *Choanephora conjuncta* serve as ß-carotene producers widely used in the industrial production of this pigment. When different-sexed strains of these fungi are co-cultured on specially selected media, the carotene yield is about 3-4 g/litre of medium.

The production of ß-carotia by *B. tnspora* is carried out in complex media, such as corn-soya media containing vegetable oils, paraffin, surfactants and some special stimulants. In order to save money, the second products of waste - corn extract and hydrol - are beginning to be used. As stimulators of carotene synthesis ß-ionone is used, which can be replaced by cheaper citrus pulp and citrus molasses. Isoprene dimers or trimers, as well as cyclohexane, cyclohexanone and trimethyl derivatives, among which 2,6,6-trimethyl-1-acetylcy-clohexane (TAC) is the most effective. Pyrrolidone, succinimide, nembutal and isoniazid can also be activators of carotinogenesis in *B. tnspora*. The addition of these activators, especially the latter, against the background of ß-ionone or TAC action allows a significant increase in carotenoid yield. The stimulants are added to the culture of the producer after the end of the peri- ode of intensive growth.

The process of ß-carotene production by the fungus B. *tnspora* is multistage. According to one of the methods, first, (+)- and (-)- strains of the fungus are grown separately. The next stage is co-cultivation of the different-sexed strains in a fermenter at 26 ^{0}C and sufficiently intensive aeration. The third stage of cultivation is the introduction of a mixed culture of *B. tnspora* into a large fermenter and incubation for 6-7 days at the same temperature and aeration.

With the help of appropriate stimulants, it is possible not only to increase significantly the yield of ß-carotene, but also to change the carotenoid composition in *B. trispora*. Under the influence of some pyridine derivatives (2-aminopyridine, 4-aminopyridine), lycopene becomes the predominant pigment instead of ß-carotene, the yield of which can be more than 60 % of all carotenoids synthesised by *B. tnspora*. When 4-aminopyridine is added, γ-carotene is formed along with lycopene, and both carotenoids are synthesised in almost equal amounts. Thus, *B. tnspora* is of practical interest not only because it is one of the most active producers of ß-carotene, but also because other carotenoids (γ-carotene, lycopene, a-carotene, ß-zeacarotene) can be obtained from this organism, which also have practical applications.

Studies are being conducted to further reduce the cost of carotenoids produced by microbiological methods. It has been shown, for example, that carotenoid synthesis in *B. tnspora* can be increased almost sevenfold if the carbon source in the medium is cellobiose. In order to reduce the cost of production, it is possible to use the waste left over from the production of cellulosic materials for this purpose. On such media *B. tnspora* synthesises besides ß-carotene such practically important enzyme as ß-glucosidase. Obtaining two valuable products at the same time significantly reduces the cost of ß-carotene production.

A highly productive mutant of yeast *Rhodospondtum diobovatum* was also proposed for practical application. A carotene-containing protein preparation was produced on the basis of this strain. A method for production of carotene-containing preparation using a highly active mutant of *Mycobacterium rubrum* was also developed; *the* preparation contains α-, β- and γ-carotines, lycopene and xanthophylls (lutein, torulin, etc.). To produce xanthophylls, *the* fungus *Dacrymyces deliquescens* is used, cultivated on a medium containing glucose, glycerol and corn extract under intensive illumination. The yield of xanthophylls in this case can be about 40 mgna1L of medium.

Chapter 11

PRODUCTION OF ORGANIC ACIDS

Organic acids are widely used in the food and pharmaceutical industries, in technology and as chemical raw materials. Some organic acids (citric acid, malic acid) can be obtained by extraction from natural raw materials; others (acetic acid, lactic acid) can be obtained by organic synthesis. More than 50 organic acids can be synthesised on the basis of microbiological synthesis. By now, biotechnological methods of their production have been developed in detail. It is believed that organic acids obtained as a result of microbiological synthesis are more preferable for human consumption than synthetic acids. Organic acids for technical needs are produced chemically; those used in the food and pharmaceutical industries are produced by biotechnological processes (production of citric, lactic, acetic, itaconic, propionic and gluconic organic acids (lactic and acetic acids are also produced chemically)).

Organic acids in the system of microbial metabolism are products of degradation of the energy source and carbon Thus, citric, isolimonic, ketoglutaric, succinic, fumaric, and malic acids are intermediates of the tricarboxylic acid cycle in most aerobic microorganisms. Gluconic, ketoglucic and tartaric acids are intermediate products of direct oxidation of glucose (without phosphorylation) in some aerobic bacteria and fungi. Lactic, butyric and propionic acids are end products of carbohydrate metabolism in anaerobic bacteria. Acetic acid is a product of ethanol oxidation, and aliphatic mono- and dicarboxylic acids are intermediate products of the oxidation of normal alkanes.

Thus, the possibilities of microorganisms to produce organic acids on the basis of their metabolism are great. Selective, strictly defined conditions are required for the supersynthesis of individual acids. In the case of balanced growth of microorganisms on a complete medium, the accumulation of organic acids does not occur, because, being intermediate products in the system of microbial metabolism, organic acids are the starting material for the synthesis of other macromolecules. The time of maximum rate of organic acid formation in the cell, as well as many other metabolites, does not coincide with the time of cell reproduction rate and biomass accumulation. Supersynthesis of organic acids is observed when the growth rate of the producer decreases and the biosynthesis processes that require acids as substrates are blocked, i.e. when the dissimilation of the available endogenous substrate by the processes of synthesis of the main (nitrogen-containing) components of the cell is disturbed. Such conditions are provided, as a rule, at full or excessive content of carbon and energy source in the medium and deficiency of biogenic elements limiting cell growth. Most organic acids are produced by lim- mting the growth of producer cells with a deficit of nitrogen or phosphorus and an excess of carbon-containing substrate, so microbiological processes of organic acid production are two-phase. At the first stage, the so-called balanced growth occurs with maximum biomass accumulation and consumption of carbon and energy substrate, as well as limiting biogen; at the second stage, the cell growth rate slows down. As a result, biomass growth stops and intensive acid formation begins. The duration of the phase of intensive acid formation is determined by the presence of carbon-containing substrate in the medium. An important condition for the formation of most organic acids (except lactic acid) is a good aeration regime, as well as the pH of the medium.

The ability to produce a particular acid is a widespread property among microorganisms. Specially selected strains producing the target acid as a monoproduct with high yields and efficient assimilation of the carbon substrate are used as derivative cultures. In myogenic organic acid production, the economic coefficient for carbon reaches up to % and higher. Bacterial, yeast and fungal cultures (*Lactobacillus, Arthrobacter, Alcaligenes, Candida, Aspergillus, Penicillium, Trichoderma*) serve as producers. Fermentation methods in microbiological processes of organic acid production are diverse.

These include surface liquid and solid-phase processes, as well as deep, including flow-through cultures. In recent years, fundamentally new and efficient biotechnologies using immobilised whole cells and enzymes have been developed. The substrates used in the production of organic acids are also diverse. Glucose and sucrose used in the early 20th century were eventually replaced by more affordable complex media (molasses, hydrolytic starch). In the 1960s, new processes were developed for the production of organic acids using liquid petroleum paraffins

At present, a number of organic acids are produced on an industrial scale by biotechnological methods. Of these, citric, gluconic, ketogluconic and itaconic acids are produced only by microbiological methods; lactic, salicylic and acetic acids are produced both chemically and microbiologically, and malic acid is produced by chemical and enzymatic methods. Acetic acid is produced by *Acetobacter* and *Gluconobacter,* citric acid by *Aspergillus niger, Aspergillus wentii,* lactic acid by *LactobaciUusdelbrueckii.*

Grape wine, beer must, honey, juices of various fruits and berries after alcoholic fermentation are used as raw materials for the production of food vinegar, and an aqueous solution of ethyl alcohol is used for white vinegar. In addition to alcohol, the medium contains acetic acid and mineral salts, which include nitrogen, phosphorus, sulphur, manganese, potassium. Sometimes sources of vitamins are added in the form of various extracts. Alcohol serves as a source of carbon and energy for the bacterium.

Acetic acid was the first microbiological product produced using immobilised cells. This method can be continuous or batch. Adsorption of acetic acid bacteria on wood chips, charcoal, coke and other substrates has been used for a long time. Passing ethanol solution through generators with immobilised bacteria, 10-15% acetic acid solution is obtained. In this case, from 100 litres of anhydrous alcohol is theoretically formed^ litres of acetic acid. In practice, the yield of acetic acid from UOl of ethanol rarely exceeds 90 litres, which is due to per- reoxidation and incomplete oxidation of ethanol by bacteria, as well as segoevaporation.

Vinegar obtained by fermentation, has a pleasant aroma and flavour, which cause by-products of fermentation: esters (ethyl acetate and otherO.higher alcohols, organic acids.

Table vinegar contains from 5 to 9 % acetic acid. Vinegar with an acid concentration of 20 to 30 % is obtained by freezing the initial solution, by distillation - 70-80 % acetic acid, called acetic essence. Glacial acetic acid contains 98.0 to 99.8 per cent of the acid.

Acetic acid, or vinegar, is widely used in the food industry. Vinegar synthesised by microbiological means (food acetic acid table vinegar) is divided into grades depending on the nature of the digested substrate Izvestey apple, grape, pear and other grades of vinegar. It is also used for dissolving organic dyes, in the production of medicines, plastics, etc.

Citric acid is widespread in nature, relatively much of it is contained in some berries, fruits, especially in citrus fruits (in lemon from 5 to 10%), in leaves and stems of some plants.

Previously, citric acid was extracted in the form of calcium citric acid from the products of cotton leaf processing, fly ash stalks, spruce needles and in significant quantities from lemon fruits. However, this production is extremely expensive and small in volume, so citric acid was a scarce and expensive product.

At present citric acid is one of the main products of microbial synthesis in terms of production, its total output in different countries reaches 400 thousand tonnes per year.

Microscopic fungi of the genera *Aspergillus, Pemcillium, Mucor, Ustma,* etc. are used to produce citric acid. The main producers of citric acid today are various strains of *Aspergfilusniger* fungi.

Citric acid supersynthesis occurs when the growth of fungus-producing fungi is limited by mineral components of the medium and at the same time the carbon source is excessive. In conditions of limitation of fungus growth due to iron and manganese deficiency, after complete absorption of the deficient element from the medium, it stops growing, but continues to consume the carbon source available in the medium. In this case, the cells of the fungus begins to accumulate citric acid, which is subsequently released into the medium.

Seed in the form of spores (conidia) is grown on molasses medium by surface or deep cultivation.

Citric acid is extracted from the culture fluid as a solid soluble salt, calcium citrate, which is formed by the addition of chalk. The conversion of citric acid to the free state is achieved by the addition of a strictly defined amount of sulphuric acid:

$$\text{calcium citrate} + H_2SO_4 = \text{citrate} + \text{gypsum}.$$

The gypsum is removed by filtration. The citric acid solution is clarified with activated charcoal, yevaporated, crystallised.

Citric acid is used in the confectionery industry to acidify caramel, marshmallows, waffles, as it brings out the fruit flavour well. This organic acid is added to ice cream, food concentrates, margarine, sausages and cheese for acidification purposes.

Citric acid is used to inhibit the formation of me-lanoids in condensed milk with sugar its solution is washed and deodorised fat raw materials, treated before cold storage fresh meat, fish, fruit in order to stabilise their colour, taste and smell. Salts of citric acid are used for the manufacture of shampoos and other detergents, as they stimulate foaming and provide mechanical stability of the foam.

Lactic acid has been industrially produced by lactic acid bacteria since 1881. Only homofermentative lactic acid bacteria are suitable for the industrial production of lactic acid, producing up to 98 % lactic acid. The strains used are *Lactobacillus ddbcucck.i, Lactobacillus bulgari*™.

Lactic acid bacteria convert a wide variety of carbohydrates into lactic acid, so molasses, whey, glucose, maltose, sucrose, lactose, saccharified starch, etc. are used for industrial production of this acid.

Lactic acid is produced under industrial conditions by anaerobic deep fermentation. During fermentation, the pH of the medium is maintained by adding chalk. After 6-7 days of cultivation, 0.5 to 0.1 % of sugars and 11 to 14 % of calcium lactate remain in the medium. From 100 sugars 80 to 90

calcium lactate is obtained.
The chalk precipitate and colloids are separated by filtration. The filtrate is evaporated, cooled and crystallised. Lactate crystals are separated by centrifugation. Lactic acid is obtained from lactate by decomposition with sulphuric acid. The reaction proceeds at temperatures between 60 and 70 ^{O}C according to the equation

$$Ca(C_3H_5O_3)_{(2)} + H_{(2)}SO_4 \quad 2C_3H_6O_3 + CaSO_4.$$

The lactic acid is treated with activated carbon, filtered and packed. The end product is a liquid concentrate of lactic acid
Lactic acid is used in the preparation of jams, where it contributes to a good consistency. Lactic acid regulates the pH, improves the flavour of products and is used in the manufacture of 150 ounces of lactic acid.
Lactic acid and lactates increase the volume of bakery products and improve the quality of bread crust when low quality flour is used. Lactic acid and lactates increase the volume of bakery products and improve the quality of bread crust when using low-quality flour. The ability of lactates to retain moisture is used in the manufacture of sausages, cheese and baby food. Lactic acid is also used for faster production of milk-protein clot in cottage cheese production.

11.1. Production of citric acid

Citric acid is a tri-basic oxoacid widely found in fruits and berries. It is widely used in the food industry in the production of confectionery and beverages, in the pharmaceutical, chemical and textile industries. Citric acid was identified as a metabolic product of mould fungi in 1893 by S. Weyer. Currently, it ranks first among all organic acids in terms of production volumes (over 350,000 tonnes/year).
In microorganisms, the synthesis of citric acid is realised in the dicarboxylic acid cycle and is carried out as a result of condensation of an acid with four carbon atoms and two carboxyl groups and an acid with one carboxyl group. The pyruvic acid formed as a result of glycolysis is bound by suglecannsloton; the oxalic-acetic acid reduced in this process reacts with acetic acid to form citric acid, i.e. the formation of citric acid involves the reactions of glycolysis and a number of reactions of the Krebs cycle. At each turn of the cycle a molecule of oxalic-acetic acid interacts with acetic acid to form citric acid
The production of citric acid by fermentation of mould fungi is one of the oldest biotechnological processes (late 19th century). Improvement of the citric acid production procedure is closely related to the development of many fundamental aspects of microbiology (control of microbial

contamination of the production culture, optimisation of the composition of nutrient media, selection of high-product strains, etc.).

In the industrial production of citric acid, *Aspergillus niger* is mainly used as a producer, but *A. wentii* is also used. The fermentation process is rather complicated, as citric acid is a product of the primary metabolism of fungi, and even a small release of it into the environment indicates a pronounced imbalance in cellular metabolism. Producer growth and acid synthesis are usually regulated by the composition of the medium (sugars, P, Mn, Se, Zn). The supersynthesis of citric acid occurs at high concentrations of sugars in the medium (14-24%) and is a response of the producer to the deficiency of phosphorus, as well as other metals, although their role is not completely clear: it seems to be both suppression of anabolism and influence on surface properties and morphology. The optimum pH at the stage of acid- formation is 1.7-2.0. In a more alkaline medium the process shifts towards the accumulation of oxalic and illuconic acids. The medium is usually based on glucose syrup, starch hydrolysates or molasses, which is pre-diluted to the required level of sugars and treated to reduce the metal content. The source of nitrogen are ammonium salts (0.2 %), phosphate concentration (0.01-0.2 %). Natural oils with a high content of fatty acids are used as defoamers. The level of aeration of the culture is essential.

Several methods are used in the production of citric acid [6] The surface method is carried out on a solid bulk medium and in the liquid phase. In liquid-phase surface-fermentation nutrient medium is poured into cuvettes in a layer of 8 to 18 cm. The cuvettes are placed on racks in the fermentation chamber, previously sterilised with formalin vapours. Through special air ducts with sterile air current, the surface of the medium is sown with the initial museum culture. The seed material used is preliminarily obtained also under surface culture conditions and dried η

spores (conidia) at the rate of 50-75mg conidia per 1 m of cuvette area. Several variants of the process are known: shiftless, shiftless with refills and the film method. In the shiftless mode, the process is carried out on a single medium from the moment of spore seeding to the end of the acid formation stage.

When using the film method, 7 days after the completion of acid formation, the fermented molasses solution is drained from the cuvettes, mycelia are washed with sterile water and a new epe- duce is poured into the cuvettes. The changeless method is characterised by fractional additions of molasses under the mushroom film at the stage of acid formation (30-35% of the initial volume) - a mode with substrate feeding. This allows to increase the yield of citric acid by 15-20 % from the unit of surface with reduction of sugars consumption by 10-15 % in comparison with other- 152 u

In the first stage of fermentation (first 24-36 h), intensive mycelial growth occurs. In the fermentation stage, intensive growth of mycelium occurs at the first stage (first 24-36 h). The medium temperature during this period is stabilised at 32...34 °C, aeration intensity is 3-4 m/h of air. During active cotyledon formation the air supply is increased 5-fold. As a result of more intensive thermogenesis, the temperature is reduced to 30...32 ^{O}C. As the rate of acid formation process decreases, the aeration regime becomes less intensive. . The process is monitored by titratable acidity of the medium. The process is considered complete when the residual concentration of sugars is about 1-2 % and the titratable acidity level is 12-20 %. The citric acid content reaches 94-98 % of the level of all acids. The fermented liquor is poured into a collection tank and sent for processing and the washed citric acid is fed to the feed mill.

Solid-phase fermentation has much in common with the surface-liquid phase process. The Kojn process developed in Japan uses porous material (bagasse, potatoes, sugar beet pulp, wheat bran) as a medium. The material is sterilised beforehand, and after cooling it is inoculated with a suspension of spores. Fermentation takes place in trays at 25...30 C for 6-7 days. The formed citric acid is extracted with water. In Japan, 20 % of the total volume of this acid is obtained by the Kodai method.

Since 1950, industrial processes of citric acid production have been transferred to the conditions of deep culture. A stable process is possible if it is organised in two stages: the first stage is mycelial growth on a complete medium, and the second stage (in the absence of phosphorus in the medium) is the formation of citric acid. Deep fermentation is carried out in apparatuses with a capacity of 50 m and filled to 70-75 %. The seed is mycelium, which is also grown under deep culture conditions. In the production apparatus, where the grown mycelium is transferred through a sterile line, the nutrient medium contains 12-15 % sugars. Fermentation is carried out at 31.32 C with continuous stirring. During the process of cotyledon formation (5-7 days), the nutrient medium is realised o

intensive aeration regime (up to 800-1000 $_{m}$/h) with fractional addition of sugars, 2-3 feedings. The yield of citric acid is from 5 to 12 %, residual concentration of sugars - 0.2-1.5 %, the proportion of citrate - 80-98 % of the sum of all organic acids. In the 60s of the XX century began to develop methods of citric acid production on the basis of liquid hydrocarbons (C_9-C_3"), using yeast (*Candida*) and bacteria (*Brevibactenum, Corynebactenum,*

Arthrobacter), as well as the method of flow cultures. These technologies have not yet been realised
In the future, there are some promising prospects.
The final product is highly purified. Crystalline citric acid is obtained during the post-fermentation stage. In addition to the target acid, fermented solutions also contain gluconic and oxalic acids, residues of non-fermentable sugars and mineral salts. To isolate citric acid from this solution, it is bound with calcium hydroxide to form difficultly soluble citrate-calcium.
At the same time calcium salts of gluconic and oxalic acids, calcium gluconate ($C_{12}H_{22}CaO_{14}$) and calcium oxalate (CaC_2O_4) are formed. The calcium salts of citric and oxalic acids precipitate out, while calcium gluconate and most of the organic and mineral components of molasses remain in solution. The precipitate is separated on a vacuum filter, washed and dried Further, to transfer citric acid into a free state and release from calcium oxalate, the precipitate is treated with sulphuric acid with subsequent filtration. The citric acid solution is filtered, concentrated va- cuum-vapourconcentrated and then subjected to crystallisation under slow cooling to 8-10C. The obtained crystals are separated in a centrifuge from the mother lode and dried in pneumatic dryers at 30...35 ^{O}C. The finished product contains not less than 99.5 % of citric acid (in terms of monohydrate), ash content is not higher than 0.1-0.35 %.

11.2. Production of lactic acid

Lactic tmcnora is an organic mono-basic acid formed as a result of anaerobic transformation of carbohydrates by molecular bacteria.^ In 1847, S. Blodno proved that this acid is a product of fermentation, and L. Pasteur established that this process is caused by bacteria. The formation of lactic acid from glucose is possible in several ways. The first way is digestion by homo-fermentative lactic acid bacteria:

$C_3H_{12}O_3 \rightarrow 2CH_2OH_2CHOHCHO$ (глицеральдегид) $\rightarrow$
$\rightarrow 2CH_3COCHO$ (метилглиоксаль) + $2H_2O$;

CH_3COCHO (метилглиоксаль) + $H_2O \rightarrow$
$\rightarrow CH_3CHOHCOOH$ (молочная кислота).

$C_3H_{12}O_3$ $2CH_{(2)}OH_2OH_2OHNOSNO$ (glyceraldehyde)
2CH3COSNO (methylglyoxal) + $2H_2O$;
CH3COSNO (methylglyoxal) + H_2O
CH3H3SNONSON (lactic acid).

The second pathway, heteroenzymatic, involves the breakdown of glucose to pyruvic acid and the reduction of the latter to lactic acid:

$C_6H_2O_6 \rightarrow CH_3COCOOH + H_2 \rightarrow CH_3CHOHCOOH$.

For the industrial production of lactic acid, homofermenting lactic acid bacteria are used. In this type of bacteria, only 3 % of the substrate is converted into cellular material and the rest is transformed into lactic acid, the yield of which reaches up to 1.5 %. Theoretically, 2 moles of pactate should be formed from 1 mole of glucose. In practice, this value is somewhat lower - 1.8 moles, i.e. the yield of the product from the substrate reaches 90 %. Lactic acid is used in the food industry to produce beverages, marmalades, in canning and in fodder production. Its salts are used in pharmaceuticals. Industrial production of lactic acid began in the late XIX century with the participation of lactic acid bacteria *Lactobacillus delbrueckü, L. leichmannü, L. bulgaricus*. Lactic acid fermentation takes place in anaerobic conditions, but lactobacilli are facultative anaerobes, so during fermentation air is not completely removed from fermenters. The raw materials are sugarcane molasses and starch hydrolysates, and the concentration of sugars in the initial medium, depending on the nature of fermentation, is about 5 to 20 %. Reduced forms of nitrogen, ammonium sulphates or phosphates are used, as well as malt and maize extract as a source of growth factors It is possible to use sulphite liquor with the bacterium *L. delbruecku.* Fermentation is carried out in a deep culture at pH 6.3-6.5 and a strictly constant temperature of 50 ^{O}C. The duration of the fermentation process is up to 7-11 hours. The duration of the process is up to 7-11 days. During the fermentation process, chalk is added to the culture 3-4 times per day to correct the changing pH. The final concentration of calcium lactate formed is 10-15 %, the residual concentration of sugars is 0.5-0.7 %. At the stage of obtaining the finished product, the culture medium is heated to 80...90 °C, then neutralised with soda lime to a slightly alkaline reaction. After settling for 3-5 h the suspended particles are decanted. Then the solution of calcium lactate is fed to the filter-press. Fnltrate is evaporated to the concentration of 27-30 %, cooled to 25...30 ^{O}C and subjected to crystallisation. Washed calcium lactate is separated by centrifugation and split with sulphuric acid at 60...70 ($^{O)}$C. Raw lactic acid of 18-20 % concentration is evaporated in several stages in vacuum evaporators up to 70 % concentration. The filtered acid after the filter press is fed for filling with small amounts of chalk, whereby about 10 % of the acid is converted into crystalline lactate, which binds lactic acid.

11.3. Production of acetic acid

Acetic acid is widely used in food, chemical, microbiological industry, medicine Its production from alcohol-containing liquids was known more than 10 thousand years ago. In those times ancient Greeks and Romans used vinegar as a refreshing drink, and it was obtained mainly by leaving wine open. On a large scale vinegar was produced for a long time in open barrels, in which a film of bacteria floated on the surface. In the 19th century, surface methods were replaced by more efficient ones. The jet generator process was developed, and in the middle of the 20th century. - deep fermentation methods. At present, the improved Frings generator is in use. Acetic acid fermentation is based on the ability of acetic acid bacteria to oxidise alcohol with air oxygen with the participation of alcohol dehydrogenase into acetic acid:

$$CH_3CH_2OH + O_2 \rightarrow CH_3COOH + H_2O,$$

1 mole of ethanol produces 1 mole of acetic acid, and 12.4 per cent of acetic acid is obtained from 1 mole of alcohol.

This process can be realised by many bacteria, but in industrial technologies for the production of vinegar, vinegar-acid bacteria of the genus *Aeetobaeter* are used, *Gluconobacter* bacteria are also of interest. Most vinegar is produced by both surface and depth methods, using diluted alcohol. The surface process is carried out in jet gerators filled with wood chips with a volume of up to 60 metres. The initial nutrient solution with bacteria is sprayed on the surface of the chips, and it flows down, collecting in the lower part of the apparatus. The liquid is then collected and pumped back into the upper part of the apparatus. The procedure is repeated 3 or 4 times, with the result that within three days up to 90 per cent of the alcohol is transformed into acetate. This old method is more efficient in Frings generators with automatic temperature maintenance and forced air supply. This process produces up to 400 million litres of acetic acid per year. Modern industrial methods of acetic acid production are used to produce acetic acid.

ca is implemented in deep culture in special aeration devices with thermal stabilisation and mechanical defoaming system. Aeration speed is 3.4 m/h, rotor rotation is 1500 rpm, temperature is 30 ^{O}C. The initial inoculated mixture contains ethanol and acetic acid. The initial inoculated mixture contains ethanol and acetic acid of about 5 H7% respectively; the final concentration of vinegar after 1.5 days is 12-13%. The process is semi-flowing, casting and pouring. Every 30-35 h up to 60 % of the culture is replaced with fresh wort. In deep fermentation, the yield per 1 m is 10 times higher than in surface fermentation. By the early 1990s, up to 715 ml of 10 % acetic acid per year was produced by this method. An efficient continuous method of acetic acid production in a battery of successive fermenters (usually five units) has been developed and introduced. The cultivation temperature is 28 ^{O}C for *Acetobacter* and 35 ^{O}C when *Baci schutzenbachü* is used as a producer.

The best raw material for the process is zthyl alcohol obtained from potato-grain raw material at a concentration of about 10 %. The optimum pH for bacterial growth is about 3. If the acetic acid content in the culture increases above 8 %, bacterial growth slows down and stops at 12-14 %. For this reason, the process is carried out in a battery of apparatuses connected in series. The first one acts as an inoculator, so it is continuously fed with fresh epe- duy to maintain conditions optimal for the rapid formation of bacterial biomass. The culture from the first apparatus is fed into the second and further into the subsequent ones, while the transport of the culture liquid is carried out by air. In each apparatus, the fermentation conditions are stabilised according to the requirements of the fermentation process with a gradual decrease in the medium temperature from 28 ^{O}C in the first apparatus to 25 ^{O}C in the last one. The aeration regime is also varied from 0.4 to 0.15 $m^3/(m(^{3)}/min)$. The alcohol concentration from the second to the fourth apparatus is stabilised at the required level by feeding them medium with 40% ethanol. From the last apparatus the culture liquid with an acetate content of at least 9.0 % and at most 9.3 % is discharged.

The yield of the acid is up to 90 kg from about 100 litres of anhydrous alcohol. At the postfermentation stage after separation of bacterial biomass vinegar solution is filtered, freeing it from coloured and suspended particles, and then subjected to pasteurisationIn order to increase the concentration, the initial solutions are frozen to 20-30 %. Further concentration to obtain glacial acetic acid (98.099.8 %) is carried out by distillation.

11.4. Production of propionic acid

Propionic acid is synthesised by *propionibacterium propionibacterium* (*Propionibacterium*), which is used in the chemical-pharmaceutical industry, in the production of cosmetics and as a fungicide for grain preservation. The chemistry of propionic acid formation is as follows: pyruvic acid is converted to oxalic-acetic acid through malic and fumaric acids to succinic acid. Succinic acid with the participation of ATP and CoA is converted into succnyl-CoA, the latter under the influence of methylmalonyl-CoA isomerase and with the participation of coenzyme B is converted into methylmalonyl-CoA. As a result of carboxenylation, methylmalonyl-CoA is cleaved to form free CoA and propionic acid. Among industrial strains-producers are the bacteria *Pr. s.* sabermosum. *Arabmosum, Pr. shermami, Pr. rubrum* and others. Bacteria use various sugars (lactose, glucose, maltose, sucrose, organic acids such as malic acid and malic acid) as fermentation substrates.

Propionic acid is produced in deep aerobic culture on media containing:

sugars - 2 %, organic nitrogen - 0.4 % (source - yeast extract), as well as salts of lactic acid. The process is realised in 12 days at 30 ^{O}C and pH 6.8-7.2; more than 70 % of sugars are transformed into organic acids and less than 20 % of carbon substrate is used for carbon dioxide formation.

11.5. Preparation of itaconic acid

Itaconic acid is an unsaturated bis-basic acid. Its formation by mould fungi was discovered in 1931 by Kinoshita. This acid is an important intermediate for the production of polymers. It forms copolymers with esters and other monomers, so it is used in the production of synthetic fibres and resins, a number of adhesives, surfactants, dyes and other complex organic compounds.

The synthesis of itaconic acid is related to Krebs cycle reactions. Its initial product is cisaconitic acid, which upon decarboxylation, as a result of electron transfer of the double bond from position 2,3 to position 3,4, is converted into itaconic acid.

Itaconic acid is obtained by surface and deep fermentation methods. Selected fungal strains (*Aspergillus itaconicus, Asp. terreus*) are used as producers. The process is similar to that of citric acid production. The media contain high concentrations of sugars, usually molasses is used in case of phosphorus and iron deficiency. The peculiarity of the process of obtaining this acid is the high demand of the producer for zinc, magnesium and copper salts. At surface fermentation during 10-12 days about 60 % of the product in terms of sugar is formed, the share of the target acid in the mixture (succinic, oxalic and fumaric acids are also synthesised) - over 90 %. The content of itaconic acid reaches 15-20 %, residual concentration of sugars does not exceed 0.6 %. Unlike citric acid, itaconic acid is a toxic product; at its concentration of about 7 %, the growth of the producer is inhibited and the rate of acid production decreases. The toxicity of itaconic acid is neutralised by fractional additions of hydroxyammonium, the pH of the medium being stabilised at 3.5-3.8. In deep fermentation, the final concentration of itaconic acid is lower than 4-6 %. The marketable product is 92 % crystalline itaconic acid, the rest is moisture (3-6 %) and other acids (1-3 %).

11.6. Production of gluconic acid

Gluconic acid is a mono-basic pentanoic acid obtained by enzymatic oxidation of glucose with the participation of gluco-zoo oxidase. Gluconic acid has many applications. Sodium gluconate, a complexing agent with metals, is used in the production of detergents; calcium, iron and potassium salts of gluconic acid are used in medicine and food industry.

Gluconic acid is produced by fungi (*Pemcillium, Aspergillus).* Fermentation

on an industrial scale is carried out by surface and deep fermentation methods; media with high (up to 30-35%) content are used.
The process is completed at a residual sugar concentration of about 1 %, with magnesium sulphate, potassium phosphate, nitrogen source and calcium carbonate in the media. The process is completed at a residual sugar concentration of about 1 %. The finished product is crystalline salts - gluconates.

11.7. Production of fumaric acid

Fumaric acid is used in the production of syi- thetic resins, inks, varnishes, and the resins are used to make printing inks. Magnesium and sodium salts of fumaric acid are used ВМе?(icine)-.
Fumaric acid, a metabolite of the tricarboxylic acid cycle, is present in all living cells, but is rarely excreted into the environment. Various tribes (*Penidlliurn, Aspergillus, Rhizopus*) are producers of this acid. *Rhizopus* is the most active. Media for fumaric acid production contain glucose in concentration of 5-10 %, limiting factor - nitrogen, zinc. Fermentation takes place in conditions of intensive aeration by surface or deep method. In the course of fermentation, the medium is neutralised with calcium carbonate or alkali solution. The maximum yield of acid is 58 % of the consumed glucose. Biotechnological methods for the production of organic acids are improving. Recently in Japan, a method for the production of 2-ketogluconic acid based on the biosynthesis *of Pseudomonas* bacteria has been developed*; the* acid yield reaches 90% of the sugar used. *A* technology for the production of oxalic acid on media with sugars on the basis of fungi *A. ozyzae* was developed, and on the basis of selected yeast strains (*Candidalipolytica)* technologies for the production of citric and isolimonoic acids were created. Specially selected strains of yeasts of the genus *Candida* synthesise fumaric, malic, succinic acids on media with normal paraffins. The process on the given raw materials of constant composition is more stable than on complex natural media based on molasses; the stage of isolation and purification of the finished product is simplified.

Chapter 12

SOURCES OF LIPIDS AND THE MAIN METHODS OF THEIR ISOLATION

Lipids are a large group of natural substances, different in chemical structure and physicochemical properties. There are several interpretations of the concept "lipids" and various schemes of their classification based on the properties of these substances The general property of lipid compounds is the ability to dissolve in ether, chloroform and other organic solvents (but not in water).

Lipids can be divided into two major groups according to their structure:

- simple lipids, or neutral fats, represented in most organisms by acylglycerols, i.e. glycerol esters of fatty acids (free fatty acids occur in cells only as a minor component);
- complex lipids containing phosphoric acid in a mono- or diether bond are phospholipids, which include glycerophospholipids and sphingolipids. Complex lipids include compounds linked by a glycosidic bond to one or more monosaccharide residues, or glycolipids, as well as compounds of steroidal and isoprenoid nature, including carotenoids.

Until the 1920s, lipids, especially neutral lipids, were regarded only as a reserve material that could be replaced by other substances of equal calorific value without much damage to the organism's vital activity. The first evidence that lipids contain compounds physiologically necessary for higher animals was obtained in 1926 by the Dutch researchers Iwaisomi Boer. A little later it was established that for most living organisms these compounds are polyunsaturated fatty acids (linoleic, linolenic and arachidonic acids) - vitamin F.

It was subsequently found that lipids fulfil a variety of biological functions in microbial cells. They are part of such important structures as the cell membrane, mitochondria, chloroplasts, and other organelles. Lipoprotein complexes play an important role in metabolic processes. They are largely responsible for the active transport of various substances across boundary membranes and the distribution of these substances within the cell. Such properties of organisms, as thermotolerativity and thermophilicity, psychrophilicity, acid resistance, virulence, resistance to ionising radiation and other characteristics are largely related to the composition of lipids. In addition, lipids can fulfil the function of reserve products. These include poly-₽-hydroxybutyric acid dumpling, formed by many bacteria, and acylglycerols, particularly triacylglycerol, which is accumulated in large quantities by some yeasts and other fungi.

12.1 Industrial production of lipids

The issues of industrial production of lipids by microorganisms are given close attention both in our country and abroad. Microorganisms can be used to produce phospholipids, glycolipids, essential fatty acids EFAs and preparations". their axiov^yyneeded in medicine, medicine, agriculture, food and other industries.

A number of yeasts and mycelial fungi are considered as potentiated producers of lipids, including lipid analogues of some types of vegetable oils. In the world practice there are no production facilities with the purpose of microbial lipids production. However, changes in the world market conditions do not exclude the expediency of lipid extraction by microbiological synthesis.

Currently, only yeasts are used to produce lipids in small quantities, and lipids are a by-product of the main production (in the production of protein-vitamin concentrates from petroleum hydrocarbons). The production of lipids from mycelial fungi, as well as bacteria, algae and protozoa has not yet gone beyond laboratory studies. One of the reasons for the slow resolution of the issues of bacterial lipid production is the presence of compounds toxic to the macroorganism.

Yeast can produce lipids using various substrates: hydrolysates of plant raw materials, sulfite liquors, petroleum hydrocarbons, and others. The efficiency of yeast fat production is largely related to the amount of basic raw materials required to obtain a certain unit of yeast mass, and its cost. In addition, the raw materials, on the basis of which the nutrient substrate for yeast growth is prepared, should provide lipids that meet the requirements of the industry, which processes lipids into various products.

The most developed technological schemes of lipid production with the help of yeast on hydrolysates of peat of low stage of decomposition and petroleum hydrocarbons. These schemes differ in the fact that when lipids are produced on peat hydrolysates, yeast fat is the main product, and when hydrocarbons are used, yeast fat is a by-product formed as a result of yeast biomass purification from residual hydrocarbons. In this connection, the fractional composition of lipids obtained by these ways is very different: the dominant fraction of hydrocarbon yeast is phospholipids, the main fraction when obtaining lipids on peat hydrolysates is triacyl-glycerols. For example, in the Russian Federation, yeast lipids are produced at the Kstovo pilot plant of protein-vitamin concentrates, which produces hundreds of tonnes of this biosnitrogen product. In the next few years, it is planned to commission several plants to produce lipids from yeast by a method similar to the Kstovo method.

The method of lipid production from hydrolysates of low decomposition peat includes several main operations: obtaining of gndrolysate-forming agent, furfural blowing and neutralisation of hydrolysate to pH 5.5-6.0, introduction of mineral nutrition sources into the hydrolysate, cultivation of yeast - lipid producers, separation of biomass and extraction of lipids from it. Consequently, the whole process is similar to the method of fodder yeast production, except for additional operations related to lipid extraction. The solvent system used for this purpose is identical to that used in the oil and fat industry. The biomass remaining after lipid extraction (bio meal) is suitable for feeding agricultural animals.

Lipid producers on peat hydrolysates are *Ltpomyees hpoferus* strains isolated from the Institute of Microbiology of the National Academy of Sciences of Belarus, whose biomass contains up to 40 % of lipids and more. From one tonne of absolutely dry peat it is possible to obtain 50-70 kg of yeast lipids containing up to 70-75 % of triacylglycerols.

In addition to peat hydrolysates, other hydrolysis media, such as wood hydrolysates or mixed substrates of wood and peat, can be used for cultivation of lipid-forming yeast and lipid production according to the above scheme.

12.2 Practical applications of lipids

Through numerous experiments it has been established that yeast lipids and products of their processing can be used in various branches of national economy: in textile, ceramic, leather, metalworking (steel sheet rolling, wire drawing, tin tinning) industries. Yeast lipids are used in the production of rubber, rubber, pharmaceuticals, cosmetics, soap, olive oil, in the process of flotation of ores and others. Finally, as experiments have shown, yeast lipids can be used in the feeding of farm animals and birds. In this case, the process of their extraction from cells is excluded from the lipid production scheme, and the fat-rich biomass of microorganisms is used for feed purposes.

After the Second World War, a considerable amount of work was directed towards the study of the possibility of obtaining microbial lipids for food purposes. Swedish researcher S. Lundin showed that yeast fat (*Rhodotocula gracúis*), rich in physiologically essential fatty acids, can be successfully used for food purposes in addition to technical ones. A diet of 25 g of yeast fat can provide the human body with 10 g of lipids, 6 g of protein and many other essential substances, which satisfies the daily requirement for these compounds by 20 per cent.

The production of microbial fat for food purposes already took place in Germany during the First World War. The nutrient medium was molasses and other sugar-containing substrates, and the producer was the yeast-like fungus *Endomycopsis vernahs*. The fat-rich biomass was consumed, from which a paste known as "Everial" or "Miceta" was prepared.

By combining nutrient media, as well as by selecting the producer and conditions of its cultivation, it is possible to obtain lipids that meet the requirements of various industries and agriculture. For example, lipids with up to 65-70% unsaturated fatty acids are favoured in poultry feed. Microbial lipids containing significant amounts of fatty acids with two double bonds can be used for the preparation of varnishes, paints, and medical preparations that help prevent atherosclerosis and thrombosis. Lipids with a predominance of saturated fatty acids can be used in the production of technical lubricants. In the first case, lipids of mycelial fungi and yeast *Lipomyces hpoferus* meet such requirements, and in the second case, *lipids* of *Candidahumicola* grown ongndrolysate wood meet such requirements.

To summarise, it should be noted that the composition of lipids (and hence the area of their possible use) is largely determined by systematic position of the organism-producer. At the same time, the ratio of individual components in the composition of lipids is determined by the specificity of the raw material and the chemical conditions of cultivation. These regularities of lipidogenesis are very significant in the organisation of industrial production of microbial fat, since under specific conditions they make it possible to obtain a product of strictly defined composition and properties. Such controlled microbial synthesis can meet the requirements for lipids of various branches of the national economy.

Chapter 13

PRODUCTION OF NUCLEOTIDES AND NUCLEIC ACIDS

The microbiological synthesis of nucleotides and their derivatives is still rather limited. The fields of application of these compounds are narrow, except for nucleotides used as flavouring food additives, especially in the production of artificial food. Nucleotides and their derivatives can be used as therapeutic drugs, in laboratory biochemical practice. A characteristic feature of most methods of nucleotide production is the need to introduce a metabolic precursor into the microorganism cultivation medium or reaction mixture.

The production of adenosine diphosphate (ADP) and adenosine triphosphate (ATP) is based on the microbial culture reaction of phos- formation of adenine or adenylic acid (5'-AMP).

In the case of BrewMrtmum *arnmornagenes* ATCC 6872, this bacterium is grown on medium containing a high concentration of glucose, urea, yeast extract, phosphorus, magnesium, calcium and biotin salts. During cultivation, adenosan is added to the medium, which undergoes enzymatic phosphorylation to form adenylic acid, ADP and ATP. Under similar cultivation conditions, but with the addition of guai guaiosine, guaiic acid, guanidine diphosphate (GDP) and guanidine triphosphate (GTP) can be obtained. The phosphorylating agent appears to be ATP. It is not possible to accumulate ATP in the reaction sphere without transferring it to the acceptor molecule of the introduced adenosine.

Another source of enzyme systems that carry out phosphorylation of 5-AMP can be baker's yeast (*Saccharomyces cerevisiae*) taken as homogenate or acetone powder. The reaction mixture contains 5'-AMF, glucose, phosphate buffer (pH 7.0) and yeast prepared as described above. Following glucose depletion with minimal inorganic phosphate, a maximum of ATP accumulation is observed. After the maximum ATP is reached, its decomposition to ADP and AMP is observed with a simultaneous increase in the level of inorganic phosphate. The product yield is significantly influenced by the phosphate concentration present in the phosphate buffer. Deviation from the optimum value may lead to a number of undesirable reactions: may cause the formation of inosinic acid, inosine, and hypoxanthine. It is likely that externally introduced adenosine is necessary for the "seed" reaction. This is equally true for 5'-AMP. Hence the need for a high concentration of glucose and a balanced amount of phosphate during the reaction.

ATP synthesis occurs by a slightly different pathway when adenine (instead of adenosine) is introduced into the medium. In this case, the key reaction is

N-ribosidation of adenine. The process is usually carried out as a typical aerobic fermentation with the addition of a precursor. *Brevibactenum ammoniagen.es* or *Corynebactenum* sp. A medium with a high concentration of glucose is also required. The first step of the synthesis involves the formation of AMP from adenine introduced at the end of the exponential growth phase of the culture and phosphoribosyl pyrophosphate (PBPP) synthesised by the cell. The reaction is carried out by adenine phosphoribosyltransferase.

The further process occurs in the same way as in the presence of AMP. FRPF in cells is formed by the reaction between ribose-5-phosphate (P-5-F) and ATP mediated by ATP: *D-ribose-5-phosphate* pyro-phosphate phosphate transferase. ATP and P-5-F are formed as a consequence of glucose catabolism: P-5-F during the functioning of the GMF pathway, and ATP as a pro- duct of glycolysis.

Microbiological synthesis of nicotinamidine dinucleotide (NAD) occurs by culturing *Brevibactenum ammoniagenes* ATCC 6872 with a precursor. Cultivation of the producer is recommended on media containing glucose, urea, yeast extract, phosphate, magnesium salts, calcium-ibiotin. Precursors of NAD in the medium can be nicotinic acid and nicotinamide, which in the conditions of the experiment are usually introduced on the second day of culture growth. By the fifth day, NAD accumulates in the medium and the culture enters the phase of stationary growth. In addition to NAD, the medium contains an intermediate product of its synthesis - nicotinic acid mononucleotide. Nicotinic acid is included directly in the synthesis, and if its amide is introduced into the medium, enzymatic deamination with the formation of acid occurs. For the production of NADPH, a system using extract from *Proteus mirabilis* IFO 3849 cells as an enzyme preparation has also been proposed, and the components are NAD, n-nitrophenylphosphate, nicotinamide, zinc sulfate, and acetate buffer (pH 4.0). However, in this case, in addition to NADPH, synthesis of NAD-diphos- phataiin some NAD analogues takes place.

Inosinic acid is a metabolic precursor of the most important purine nucleotides, adenylic and guanylic acids. For this reason, accumulation of inosinic acid can occur if the producer has a block in the enzyme systems that carry out further transformations

Obviously, the structure should not be deprived of such necessary and important metabolites (adenylic and guanylic acids). Introducing them into the medium to compensate for their deficiency is possible but rather expensive, so it is preferable to introduce a nucleoside - inosine (nucleoside)

into the reaction sphere. To produce 5'-inosinic acid (5'-IMP), it is necessary to have a source of an enzyme that phosphorylates nnosn at the 5'-position of the ribose residue. Sources of phosphorylating enzyme - nucle- osid diphosphate transferase - can be many microorganisms belonging to the genera *Flavobactenum, Serrana, Staphylococcus, Pseudomonas.* Phosphorus donors in this reaction can be nucleotides, as well as non-natural compounds - *n-nitrophenyl phosphate* (*n-NPP*), benzylphospho- phatphenyl phosphate.

In the USSR, *Pseudomonas* strain was proposed as a source of phosphorylating enzyme. The culture is preliminarily grown on medium containing glucose, peptone, and yeast extract. Live cells dried under vacuum or acetone, as well as frozen cells, are used as the enzyme preparation that carries out the reaction, since the localisation of the enzyme is intracellular. Of the phosphate donors mentioned above, t-nitrophenylphosphate is considered the best. The reaction takes place in acetate buffer at pH 4.0 in the presence of copper and zinc ions, which markedly increase the yield of the product.

The most important factors determining the activity of the inosine phosphorylation reaction are the concentration of donor (phosphorus), acceptor (inosine), and the number of cells. At optimal ratios of *n-NPP* and inosine, *Ps. tnfolu* cells can phosphorylate up to 90% of inosine in the reaction mixture. The yield of inosinic acid can be increased if surfactants, in particular various tweens, are added to the reaction mixture. The concentrations of the additives are selected experimentally.

As in all other cases when microbial cells are used as enzyme preparations, it is necessary to determine the activity and conditions of its manifestation in enzymes that destroy the target product of synthesis. The presence of active phosphomonoesterase and 5'-nucleotn- dase is most undesirable for the synthesis of inosi- new acids. If it is impossible to suppress their activity by inhibitors, it is necessary to choose conditions under which it will be minimised. In the first place, the most easily regulated conditions are temperature and pH of the medium. Inosinic acid is isolated as a barium salt and putemion-exchange chromatography.

Let us also consider the *synthesis of guanosine polyphosphates*. In microorganisms, guanosine polyphosphates are found among the metabolic products of nucleotide nature, mainly tetra- or pentaphosphates. They act as regulators in numerous biochemical reactions, and therefore their production is the focus of attention of many specialists.

Guanosine polyphosphate accumulation was detected in the culture of *Brevibactenum ammoniagenes* KV 13510 in the presence of guanosine-5'-

mono-phosphate (G-5'-P) or xanthosine-5'-monophosphate (X-5'-P). The strain is a mutant that requires the presence of expensive components in the medium for industrial production: yeast extract, biotin, thiamine, pantothenic acid and others.

Xanthosine-5'-monophosphate is a metabolic precursor of gua-nilic acid formed after amidation of the purine base of glutamines in the presence of ATP.

The actual process of enzymatic conversion of metabolic precursors (G-5 -F or Ks-5 -F) to nucleoside polyphosphate occurs in a medium containing glucose, as well as high levels of phosphate and magnesium. In addition to polyphosphates, 5'-GMF, 5'-GDP, and 5'-GTP accumulate in the medium, but guanosine polyphosphates are formed in a narrower pH range and at a different temperature optimum.

The enzyme ATP-nucleotide pyrophosphate transfer enzyme, which transfers pyrophosphate residues of nsATP ^ acceptor molecules, was isolated from *Streptomyces adephospholyUcus* A-4668. The reaction products can be nucleoside-3'-diphosphate-5'-mono(di- and tri-)phosphates, such as pppGpp, ppGpp, pGpp, and adenine and inosine derivatives, such as pppApp, pppArr, and pppipp.

Chapter 14

PRODUCTION OF SUGARS AND POLYSACCHARIDES

Polysaccharides (or glycaires) are polymers constructed of at least 11 monosaccharide units. They may consist of one or more types of monosaccharides. Accordingly, homopolysaccharides and heteropolysaccharides are distinguished. Polysaccharides, obligatory components of all organisms, constitute the majority of carbohydrates found in nature, the predominant share in biomass, and therefore the bulk of organic matter on Earth.

Polysaccharides are found as independent polymers, as well as in complexes with nucleic acids, proteins, lipids, phosphate. They are characterised by their monomeric composition and structure. Polysaccharides of microorganisms are particularly diverse. Some of them are close or identical to polysaccharides of plants and animals. But the vast majority of microbial polysaccharides have a unique structure, specific to the species or to the serological group of the species. Microbial glycans often contain previously unknown monosaccharides, which are not found in animals or terrestrial organisms

The fact that mucus formed by many microorganisms may be of carbohydrate nature was known in the time of L. Pasteur. However, special attention to microbial polysaccharides began to be paid only in the early 20s of the XX century, when it was learned that the substances determining the serological specificity of pneumococci are polysaccharides. At present, the study of microbial polysaccharides has acquired special importance due to the possibility of their wide application in medicine and a number of areas of national economy.

Polysaccharides of microorganisms are divided into intracellular and extracellular polysaccharides according to their localisation. Intracellular polysaccharides usually include polysaccharides of cytoplasm, membranes, and cell walls, and extracellular polysaccharides include polysaccharides of capsules, sheaths (wall structures), and free mucus not adjacent to the cell wall. Sometimes polysaccharides localised outside the cytoplasmic membrane are also included in the extracellular group. In this case, cell wall polysaccharides are also included in the group of extracellular polysaccharides. In some microorganisms it is really difficult to distinguish the boundary between the capsule and the cell wall.

Three groups of polysaccharides are often ^localised: intra tricellular (cytoplasm, membranes, periplasm), cell wall polysaccharides and extracellular (sheath capsules and free mucus).

The term "exoglycans" is applied mainly to polysaccharides of free mucus. Sometimes exoglycans are also referred to as capsular polysaccharides.

Microbial polysaccharides are grouped according to their functions: reserve polysaccharides involved in active transport, support polysaccharides involved in cell-to-cell interaction, defence polysaccharides, and others

Some researchers classify polysaccharides by considering their topology and function. According to this, cellular polysaccharides are subdivided into two groups. One includes reserve-energy and modifiers (intracellular), the other includes structural and structural-metabolic (in the cell wall). The extracellular ones include structural-metabolic glycans and exoglycans proper, which are released during hyperproduction.

14.1. Polysaccharides of cytoplasm and membrane structures

Cytoplasmic polysaccharides are found in two forms: they can be dispersed in the cytoplasm or aggregated into granules. Usually, the cytoplasm of bacteria contains 20-30 % of polysaccharides, and under conditions favourable to their accumulation, up to 50-60 % of the dry cell mass. Homoglycans are most often found in the cytoplasm of microorganisms, of which glycogen-type glucans are especially common. They were isolated from the cytoplasm of many prokaryotic and eukaryotic microorganisms: representatives of different genera of bacteria *Agrobactenutn, Arthrobacter, Bacillus, Clostridium, Escherichia, Mycobacterium, Nostoc, Rhodopseudomonas, Rhodospirillum, Streptococcus*, as well as yeasts, mycelial fungi, ciliated and flagellated protozoa, and some algae. In addition to glycogen-like polysaccharides, starch, mannans, levans, arabans, and xylans were found in the cytoplasm of microorganisms. Glycogen and other homoglycans of the cytoplasm can form complexes with DNA, RNA, proteins, lipids, and phosphate. Heteropolysaccharides are less frequently detected in the cytoplasm. However, they are predominant in *Streptomyces* and *Mycobacterium* species. The functions of cytoplasmic polysaccharides have not been fully elucidated. Until recently, it was believed that their main or even only purpose was to be a reserve source of carbon and energy for the cell. They are consumed, for example, during endospore maturation in bacteria of the genera *Bacillus* and *Clostridium*. But now it is clear that polysaccharides of cytoplasm can fulfil a number of other important functions. It is assumed that complexes of homoglycans with other components of the cytoplasm participate in the mechanisms of cell regulation, controlling the synthesis of various substances, growth and cell division Thus, glycogenribosomal complexes control protein synthesis.

Glycogen can exert a radio-protective effect on associated nucleic acid molecules.
In the membranes of microorganisms, an average of 2 to 5 %, sometimes (in *Micrococcus luteus*) up to 15-20 % of carbohydrates from the membrane mass is found. In some cases, these carbohydrates completely or partially represent the complete material of cytoplasm or cell walls. Nevertheless, it has been reliably proved that membranes of Gram-positive and some Gram-negative bacteria contain glycolipids and glycoproteins.
All Gram-positive eubacteria, except for micrococci and some streptococci, as well as yeasts and mycelial fungi, contain teichoic acids (1-2 % of dry biomass) in the region of the cytoplasmic membrane. Teichoic acids are acidic polysaccharides of unusual structure. During their hydrolysis, along with monosaccharides, substances belonging to other classes of compounds are formed. The diversity of teichoic acids is determined mainly by the number of residual sugars present in them and the presence of bonds of various types. Membrane teichoic acids are always glycerophosphate polymers, often bound to glycolipids and phospholipids (lipo- teichoic acids). In some microorganisms, only lipoteichoic acids are detected and no free teichoic acids are present. Teichoic acids are not detected in the membranes of Gram-negative bacteria.
Membrane glycolipids participate in the biosynthesis of polysaccharides and transport of sugars. Teichoic acids apparently regulate ion exchange (bind divalent cations, which is necessary for the normal functioning of enzymes localised in the membrane), act on the binding of amino acids to tRNA, act as a link between the membrane and the cell wall, and exhibit antigenic activity.

14.2. Cell wall polysaccharides

In Gram-positive eubacteria, polysaccharides make up from 30 to 60 % of the dry mass of the cell wall. A significant part of them is part of the murein complex, the content of which in Gram-positive eubacteria reaches 50-90% of cell wall substances. The linear polysaccharide chains of murein are constructed of repeating Ᵽ-1,4- linked units of N-acetylglucosamine and N-acetylmuramic acid. Muramic acid is a derivative of glucosamine containing *D-lactic* acid.
The cell walls of some Archaebacteria that give positive Gram staining contain pseudomurein, the glycan part of which consists of N-acetylglucosamine, N-acetylgalactosamine, and N-acetylthylthalosaminuronic acid. Muramic acid? of pseudomurein has not been found. In a number of Gram-positive Archaebacteria, the cell wall is constructed only of acidic heteropolysaccharide, which includes

galactosamine, neutral sugars, and uronic acids. The cell walls of the vast majority of Gram-positive eubacteria, with the exception of Mycobacterium and Corynebacterium, contain teichoic acids. Their amount usually reaches 50-90 % of the cell wall mass, in Streptomycetes it varies from 4 to 50 %. As a rule, either glycerol teichoic acid or ribitetichoic acid is found in microorganisms. However, in *Streptococcus faecalis* and one strain of *Streptomyces* sp. both types of teichoic acids were found.

Other polysaccharides contained in the cell walls of Gram-positive bacteria are very diverse; most often they are heteroglycans. They include neutral monosaccharides, aminosaccharides, uronic acids, acetyl groups, and phosphoric acid residues.

The cell walls of Gram-negative bacteria contain from 1 to 50 % of polysaccharides. Among them, polysaccharides of the murein complex do not occupy a dominant position, as its content™ - on average only about 5 % of cell wall substances. Theichoic acids are detected only in some representatives of Gram-negative bacteria.

Lipopolysaccharides (LPS), biologically active substances involved in the formation of the outer membrane, are particularly characteristic of the walls of Gram-negative bacteria. The number of different lipopolysaccharides is large, and despite intensive study, the structure and composition of many of them are not completely known. The polysaccharide part of the complex distinguishes between a basic structure and O-specific side chains. The monosaccharide composition, the variation of bonds and their structure mainly determine the biological activity of lipopolysaccharides, which fulfil the function of somatic antigens. The carbohydrate component of lipopolysaccharides is usually a heteropolysaccharide. Neutral sugars, amino sugars, uronic acids, methyl and acetyl groups and, most characteristically, 3,6-dideoxy derivatives of sugars, which are rarely found in other natural objects. The most common of these sugars are abequose (3,*6-dideoxy-D-galactose*), colitose (3,*6-dideoxy-L-galactose*), tivelose (3,*6-dideoxy-D-mannose*), ascarilose (3,6-dideoxy-D-mannose) and paratose (3,*6-dideoxy-D-glucose*). These sugars often determine^ the serological, specificity of the bacteria. In the O-specific side-chain of the lipopolysaccharide of one of the species of the genus *Pastora**, *6-deoxy-D-mannoheptose* was found, the first 6-deoxyheptose found in nature. Monosaccharides of a special type - glycolactyl acids - were found in the composition of O-antigenic polyeaccharides of representatives of the genus *Sheila.* The polysaccharide components of LPS of a number of bacteria are characterised by complexity. They may contain up to 6 or more different unsubstituted and substituted monosaccharides. In Enterobacteriaceae, LPS molecules can form complexes with peptidoglycan, acidic capsulin-miglycans and other heteropolysaccharides of the cell.

Polysaccharides are the main components of the cell wall of yeasts and mycelial fungi. They can account for up to 90% of the cell wall mass (*Saccharomyces cerevisiae*). Homoglycans are often found in yeast: glucans, mannans, chitin (polymer of N-acetylglucose-mine). Yeast mannans often contain phosphoric acid residues and/or methyl groups, and glucans often contain acetyl and amino groups. In many yeasts, heteropolysaccharides - galactomannans and glucomannans, as well as acidic glycans constructed from 3-4 monomers - have been detected in the cell wall. Apparently, several polysaccharides are present in yeast cell walls. Yeast cell wall glycans may be associated with peptides.

Chitin is the main polysaccharide component of cell walls of most mycelial fungi studied. Chitosan, glucosamine polymer, glucans (sometimes cellulose), galactans and various heteropolysaccharides including unsubstituted and substituted sugars and uronic acids are also found.

The structural microfibrils of cell walls of most algal microforms consist of cellulose, and in some representatives - of other homopolysaccharides, often of xylans and mannans. Their quantity may reach 50-80 % of the dry cell wall mass. The matrix polysaccharides are represented mainly by heteroglycans, and sulfated polysaccharides are also found.

Cell wall polysaccharides fulfil a variety of functions, many of them determine the mechanical strength of cell walls, which is why they are often referred to as skeletal polysaccharides. Lipopolysaccharides, teichoic acids, as well as heteropolysaccharides of a number of Gram-positive bacteria are responsible for the antigenic activity of cells and LPS of a significant number of Gram-negative bacteria are toxins. LPS of Enterobacteriaceae protect cells from the inhibitory action of long-chain fatty acids, allowing these bacteria to survive in the intestines of animals. The presence of O-specific side chains of LPS is associated with the ability of shigellae to attach to the epithelial cell membrane. Many polysaccharides determine the resistance of the bacterium to lytic enzymes and phages. Polyanionic polysaccharides facilitate the transport of charged metabolites and substances from the environment out of the cell. In addition, such polysaccharides communicate a negative charge to the cell, resulting in mutual repulsion of cells, dispersing them in the medium. Loss of LPS O-lateral chains reduces cell hydrophilicity and leads to spontaneous agglutination. Cell wall polysaccharides of microorganisms growing on media with n-alkanes are usually good emulsifiers and thus favour the penetration of hydrocarbons into the cell.

14.3. Extracellular polysaccharides

Extracellular polysaccharides, as already mentioned, are found in the form of capsules and sheaths adjacent to cell walls, as well as free mucus. Capsules less than 0.2 μm thick, indistinguishable in the light microscope but clearly visible in the electron microscope, are called microcapsules. Microcapsules are usually more firmly bound to the cell wall than capsules. In many microorganisms. capsules have a definite structure and are clearly delimited from the SLISN In some bacteria, the capsular material is loose, structureless and easily detached from the cells, so the boundary between the capsule and free mucus in this case is difficult to determine. Such amorphous capsules are called mucous layers. Pouches here- (lichiye) of capsules have a complex structure. They often distinguish several layers with different structure. The amount of extracellular polysaccharides can be many times greater than the biomass of cells. Capsules, sheaths and mucus are not always composed of polysaccharides alone. In addition to glycans, they may include proteins, polypeptides, nucleic acids, lipids, which may or may not form complexes with polysaccharides. In *Sphaerotilus natans* they contain glucose, glucosamine, protein, lipid and phosphate. The sheaths of bacteria that oxidise reduced metal compounds often contain inclusions of their oxides. Sometimes non-carbohydrate polymers are the only components of capsules and mucus. For example, the capsules of some species of the genus *BaaUrn* are made of polypeptide, while the mucus of some strains of *Pseudomonas aeruginosa* consists of DNA.

Extracellular polysaccharides, capsular or free or both, are formed by many microorganisms. There is probably no group of microorganisms that does not possess this ability. However, the synthesis of extracellular polysaccharides is not an obligatory function of the cell, and it appears only under certain conditions. There are micro-organisms in which no capsule synthesis has ever been observed.

Extracellular polysaccharides of microorganisms are extremely diverse in composition and structure. To date, the composition of about 200 exoglycans has been investigated, and the primary structure and details of the structure of many of them have been established. More than 20 monosaccharides and their derivatives have been found in the composition of extracellular polysaccharides of various microorganisms. The most common are hexoses: glucose, galactose, mannose and 6-deoxyhexoses: fucose and rhamnose. Pentoses are less frequently detected: arabinose, xylose, ribose. Uronic acids are common: galacturonic, mannuronic and especially glucuronic acids.

Many contain amino sugars: glucosamine, galactosamine and mannosamine. Non-carbohydrate substituents such as pyruvate and acetate are often present in exoglycans; succinate and glycerol are also found. A number of extracellular glycans are characterised by the presence of rare monomers: 2,6- or 3,6-dideoxysaccharides, and some have previously unknown monosaccharides, such as glncolactl acids (in saprotrophic mycobacteria). Theichoic acids, phosphate and sulphate ions are sometimes found.

The extracellular polysaccharides of most bacterial species are acidic heteroglycans of diverse composition, constructed from 2-5, and sometimes 6-7 monomers, linear and branched, and having a regular structure of repeating oligosaccharide links. For example, *Xanthomonas campestris* synthesises the polyanionite xanthan, including glucose, mannose, glucuronic acid and O-acetyl group-puipiruvate.

Some bacteria produce ventral heteropolysaccharides.

Various homofilled charides, especially glucans, are very common in microorganisms, of which dextrans (a group of neutral glucans more or less similar in structure) are the best known. They can contain up to 200 thousand glucose residues, and can be linear or branched. The linear (main) chain is built with the participation of a-1,6-linkages, branching is caused by a-1,2-, a-1,3- and a-1,4-linkages. The molecular mass of dextrans varies from 12 to 600 million. The most active producers of dextrans are representatives of lactic acid bacteria *Leuconostoc rnesenterouies uL dextrarncurn.* Dextrans are also synthesised by some species of *Streptococcus (Str. sanguis, Str. mutans), Brevibacterium, Lactobacillus.* Almost every producer synthesises its own, somewhat different dextran. Some strains produce two different dextrans structures simultaneously.

Extracellular cellulose, a polysaccharide common in plants, is synthesised from bacteria by some representatives of *Pseudomonas, Zooglea, Azotobacter,* an overwhelming number of *Rhizobium* and *Agrobactenum* species*, and Acetobacterxylinum.* P-(1^3)-Glucan, called kurdlan, is formed by *Alcaligenes faecalis* var. *myxogenes* and *Rhizobium* species. A highly branched "-(1^4)-glucan with side chains attached by "-(^-links, a polymer like glycogen, a reserve polysaccharide of animals, many bacteria and yeasts, can be found in the culture fluid of *Neisseria perflava.* Nigeran - α-glucan with alternating (1^4)- and (^(-linkages - Formed by the fungus *Aspergilla riger.*

Branched fructose polymers - levans with (2^6)-linkages - are synthesised by acetic acid bacteria *Gluconobacter oxydans, Acetobacter aceti,* some species of *Pseudomonas, Erwinia* (*Herbicola* subgroup*), Aerémonos,*

Bacillus.Fructans of the inulin type (a reserve polysaccharide of plants of the complex family) with (2 >1)-links in the main chain and branches at position C6 are formed by *Str. mutans* strains.

Mannans have been found in cultures of some *Bacillus* and *Saugupe-bacterium* species, as well as in many yeasts.

Most often, microorganisms capable of "forming extracellular polysaccharides" synthesise capsules and free mucus. The monomer and monomer composition of mucus and capsules are the same in most cases.

However, it is not always possible to determine which polysaccharides are specific to a particular group of microorganisms. In some cases, phylogenetically similar bacteria synthesise extracellular glycans that are similar in composition and structure. These include, for example, bacterial alginate *of Pseudomonas aeruginosa* and *Azotobacier vinelandii, Rhizobium* curdlanum *and A. faecalis* var. *rnyxogenes,* acidic heteroglycans *of Corynebacterium* and *Arthrobacter,* dextrans *of Streptococcus* and *Lactobacillus,* and others. However, microorganisms that are far apart in taxonomic respect often form glycans of similar or identical composition. Usually, in this case, the polymers show functional similarity. For example, exoglycans of various phytopathogenic bacteria and meningitis pathogens are very similar in composition. In a number of bacteria, extracellular cellulose serves as a cementing material for the formation of cell aggregates.

It is known that not only different species of the same genus of microorganisms, but also different strains of the same species often synthesise different exopolysaccharides. For example, *E. coli* has about 70 serotypes that differ in the composition and structure of capsular polysaccharides and, accordingly, in their immunochemical properties. The exoglycans of effective and ineffective strains of nodule bacteria differ in monomer composition, and those of different strains of yeast of the genus *Lipomyces* differ in the ratio of monosaccharides. Extracellular glycans of *M-*, S-, and *R-forms of bacteria* may differ in monosaccharide composition.

There are cases when several different glycans accumulate in the culture fluid of one microorganism. For example, *L. mesenteroides* forms dextran and levan, *Serratia marcescens* has three exopolysaccharides: acidic glucoramna and containing glucuronic acid, rhamnoglucan and heptoglucan. The fungus *Aupeobasahurn (Pullulanoipullulans* forms two homoglucans: Aubasidan, a branched polymer with "-(1^4), ₽-(1^3)and ß-(l^-links, and the side chain may have from one to four glucose residues per triose link, and Pullulan, a linear glucan consisting of maltotriose and maltotetraose fragments joined by a-(1->6)-links.

Extracellular polysaccharides are not essential for microorganisms. There are species in nature that never form them. It has been shown experimentally that cells devoid of capsules are as viable as those that are encapsulated. Nevertheless, extracellular polysaccharides perform certain functions that contribute to the maintenance of conditions favourable for the life of the producer. Some of them are universal for all polysaccharides, as they are determined by the properties common to these substances, while others are specific for a certain polysaccharide due to the peculiarities of the composition and structure of the given polymer.

Extracellular polysaccharides protect cells from desiccation, destructive action of ultraviolet rays and various chemical agents, including heavy metals and drugs It has been observed that encapsulated bacteria are more resistant to chemotherapeutic agents than non-capsulated variants. Being located superficially, capsules protect cells from bacteriophages and their uptake by protozoa, prevent denaturation of cellular protein.

Many extracellular glycans are biologically active and determine the immunological properties and virulence of strains. As a rule, the thicker the capsule, the higher the virulence and pathogenicity of the bacterium. Some pathogenic bacteria deprived of capsules become avir- lent. *K-antigens* are aggressors that inhibit phagocytosis of the bacterium and thus create conditions for their multiplication.

One of the characteristic manifestations of the biological activity of polysaccharides is their ability to modify enzymes of various opra- nisms, resulting in stimulation or reduction of their activity. Exopolysaccharides of phytopathogenic bacteria are phytotoxins involved in the specific interaction of the bacterium with plant tissue.

In some cases, extracellular glycans serve as a reserve source of carbon and energy, and nitrogen-containing polysaccharides serve as a source of nitrogen for the producer. Polyanionic polysaccharides concentrate cations from the environment and facilitate their transport into the cell. Polysaccharide-lipid complexes of microorganisms growing on media with n-alkanes: pseudomonads, nocardia, corynebacterium, yeast are good emulsifiers. They reduce surface tension, increase the contact surface of hydrocarbon and water, form micelles and thus favour the penetration of hydrocarbons into the cell.

Neutral extracellular polysaccharides maintain the integrity of filamentous forms and various cell aggregations. In contrast, polysaccharides carrying a specific charge favour cell dispersion. The exoglycans of some bacteria are responsible for the attachment of cells to the surface. For example, *Str.*

mutans, which causes dental caries, attaches to tooth enamel with the help of dextran.

14.4. Use of polysaccharides

At present, microbial polysaccharides are used in various spheres of human activity: in medicine, pharmaceutical, food, chemical and textile industries, in hydrometallurgy, in oil production and in a number of other areas of national economy. At the same time, both intracellular and extracellular glycans attract the attention of researchers and practitioners, but in technical and economic terms the latter are preferable, as the scale of their production and application is much wider.

The possibility and prospects of using polysaccharides in medicine are largely determined by their biological activity- (n)™ . , ,

Many microbial polysaccharides have therapeutic and prophylactic effects: increase the body's resistance to bacterial and viral infections, are characterised by antitumour activity, promote wound healing and tissue regeneration, favourably affect the course and outcome of inflammatory processes, eliminate pain syndrome, reduce the side effects of drugs and radiotherapy. The therapeutic and protective effect of polysaccharides is determined primarily by their ability to increase non-specific immunobiological reactivity of the body, to influence various protective reactions that maintain the constancy of its internal environment. Advantages of many polnsaccharide preparations over other means that increase non-specific resistance of the organism are determined by the fact that they are free from additives that have undesirable effects on the organism. Some microbial polysaccharides have already found application in the therapeutic practice of various clinics around the world.

In our country, pyro-genal, a preparation isolated from *Salmonella typhi* and *Pseudomonas aeruginosa* cells, was successfully used to treat the consequences of trauma and nerve system conduction disorders, to prevent the formation of rough scars or posttraumatic scars. In Germany and the USA lipopolysaccharides isolated from various pathogenic bacteria were used for the same purpose. Bacterial lipopolysaccharides also have anti-radiation activity.

Prodigiosan, a heteropolysaccharide complex with lipids isolated from *Serrada rnarceseens* cells, and zymosan, a preparation from *Sacch. cerevisiae* cell membranes, consisting of glucan, glucomannan and minor amounts of teichoic acids, are used in clinics. *cerevisiae*, consisting of glucan, glucomannan and minor amounts of teichoic acids. These preparations normalise a number of shifts in immunobiological reactions, have a positive effect in the treatment of tumours, infectious and non-infectious diseases. Promising as antitumour agents are LPS of Gram-negative bacteria, tricellular glucan paramilon (astasian) of phytoflagellates■ *Astana longa,* extracellular polysaccharides of various yeasts of genera *Lipomyces, Cryptococcus, Bullera,* bacteria of genera *Alcaligenes* and *Agrobacterium.* The listed compounds are recommended for clinical trials. Prodigiosan shows antiviral activity. Modified (sulfated) polar mannan, an extracellular polysaccharide *of Rhodotorula rubra*, can be used as a means of prevention and treatment of atherosclerosis.

Polysaccharides with antigenic specificity are beginning to be used in medical practice as diagnostic tools. For example, polysaccharide preparations of pathogenic and conditionally pathogenic yeast species of *Candida* genus, *which* facilitate the diagnosis of diseases of candida nature, are among them. The possibility of using modified LPS-antigens of Salmonella in the diagnosis of salmonellosis has been shown.

Purified specific polysaccharides of meningococcal groups A and C (polymers of N-acetyl, O-acetylmannosaminephosphate and N-acetyl, O-acetylneuraminic acid, respectively) are used to produce meningococcal vaccines. Microbial polysaccharides can be the basis for the creation of artificial vaccines. This is achieved by changing their configuration or by conjugation with synthetic polyelectrolytes. Neutral dextrans with a molecular weight of about 75,000 produced by *L. mesenterotdes* are widely used in our country and abroad as substitutes for blood plasma. Pullulan, levans synthesised by *G. oxydans* and *Vas. polymyxa* are promising as plasma substitutes. *polymyxa.* Dextrans of a certain structure, as well as many other polysaccharides, are able to stimulate defence reactions of the organism. In clinics, they are used in combination with other drugs for the treatment of various abdominal cavity diseases. Dextran sulphates have an anticoagulant effect, replace heparin and can be used as an antithrombogenic agent. Quitin is also promising as an anticoagulant.

The wide application of microbial polysaccharides in pharmaceutical, perfume, food and other industries is determined by their properties: viscosity, rheological characteristics, ability to swell, interaction with certain structures. In pharmaceutics they are used as a basis for preparation of dosage forms: as softeners, emulsifiers and stabilisers of suspension, as bonding agents and leavening agents in ointments, pills, tablets. They provide long-term stability of drugs, stabilise and prolong their action. On the basis of some microbial polysaccharides (aubazidan, dextran), drugs of butadione, sulphur, sulphonamides, barium sulphate suspension for fluoroscopy, etc.

were created stable for several years. Macromolecular conjugates of modified dextrans with enzymes (streptokinase, trypsin, fibrinolysin) prolong the activity of enzymes and reduce their allergic effect. Microbial polysaccharides are used as gel-forming agents in the manufacture of cosmetic products, to create a hydrophilic buffer in creams, as swelling agents in the production of creams, shampoos, lotions. Some glycans can be used instead of sodium carboxycellulose used as a binder and biologically active component in toothpastes.

In the food industry, polysaccharides of microorganisms are used in the form of films - coatings for products, such as cheese to protect them from drying and mould, as stabilisers of ice cream, fruit juices, salad dressings, thickeners of syrups, jams, gravies, jellies and other culinary products. Xaithai is considered particularly promising in this respect. Slime-forming strains of *Streptococcus lactis* are used in Switzerland in the production of thick kefirs, sour cream and some soft cheeses. Exopolysaccharides of yeasts *of Saccharomyces* and *Ctreptococcus* genera, bacteria *of Azotobacter* and *Arthrobacter* genera can be used to improve the quality of bread. Adding them to flour during bread baking increases the gas-holding capacity of dough and improves its rheological properties. Bread baked from such dough is characterised by high specific volume, good porosity and slower staling.

As gel-forming agents, exoglycaires are used in the production of nuclear fuel, photographic and X-ray films, as substitutes for alginic acid in algae in the food, textile, pharmaceutical and paper industries (polyuronides of *Azotobacter, P. aerosa* and a number of other microorganisms), they can replace agar (heteropolysaccharides *of V.* subtilis and Ps. elodea consisting of glucose, galactose, fucose, glucuronic acid and glucose, *rhamnose*, glucuronic acid and glucose, *rhamnose*, glucuronic acid and glucose, respectively). *subtilis* and *Ps. elodea*, consisting respectively of glucose, galactose, fucose, glucuronic acid and glucose, rhamnose, glucuronic acid and O-acetyl groups).

Anionic polysaccharides (xanthan, zanflo - an extracellular hetero- glycan *of Erwinia tahitica* consisting of glucose, galactose, fucose, uronic acid and acetyl groups, etc.) stabilise and protect cationic aqueous enamel paints from drying. Some glycans, such as the heteropolysaccharide of *Corynebacterium equi* var. *mucilagenosus*, are highly viscous and can replace expensive adhesives. Xanthan sulphates are used as thickeners for adhesives. On the other hand, the ability of some polysaccharides to form surface films allows them to be used as anti-adhesives, for example when releasing casts from moulds. Dextran is also recommended as a lubricant.

Polysaccharides, the aqueous solutions of which are particularly stable at sharp temperature changes and in aggressive environments, are used in the oil and gas industry as stabilisers and structure formers of washing fluids intended for drilling oil and gas wells, and ensure more complete recovery of oil from oil-bearing formations. Since the early 2000s, more than half of the oil in the US has been produced using polysaccharides, mainly xanthan. Scleroglucan, a capsular linear neutral glucan formed by imperfect fungi, mainly of the genus *Sclerotium*, is being commercially tested. As a stabiliser of drilling clay slurries, *a* linear heteroglycan of the obligate *methylotrophic* bacterium *Methylobacillus methylophilus* consisting of glucose, galactose, mannose, rhamnose and glucuronic acid is promising. The application of polysaccharides in the oil industry is promising in technical and technical terms.

Polysaccharides of some microorganisms (*A.puUulans pull*ulan, heteropolysaccharide of bacteria of genus *Metámonos,* etc.) are flocculation agents and are used in hydrometallurgy to obtain metal components in the form of gels including its insoluble precipitates. The process is realised in the purification, separation and concentration of metals from solutions of their salts or mixtures of salts.

Sefadexes are prepared on the basis of dextrans and are widely used in laboratory practice for gel filtration. Polyanionic glycans, e.g. xanthan, chitin, are recommended for water purification from heavy metals, as well as for industrial synthesis of polymers for their extraction from organic solvents. Chitin can also be used for wastewater treatment.

Aubazidan is used as a carrier for immobilised "-amylase. Curdlan and chitin are promising for enzyme immobilisation.

Microbial levans are sources of pure fructose preparation, chitin - *D-glucosamine* and *N-acetyl-D-glucosamine* - compounds used in chemical synthesis. From yeast mannans, maiose can be obtained.

The polysaccharides of some bacteria, such as the levan *of Vas. polymyxa levan*, have proved useful in agriculture. When applied to the soil, they increase the survival rate of cultivated plant seeds by helping them to retain moisture. Alginate films are used to cover the roots and seeds of plants to prevent them from drying out during storage and transport.

The possibilities of practical application of microbial polysaccharides have not been fully explored yet. A comprehensive study of glycans in this regard opens up new perspectives and will undoubtedly lead to the expansion of the relevant area of the microbiological industry.

14.5. Industrial production of microbial polysaccharides

The expansion of the range of microbial polysaccharides of practical importance has led to advances in the organisation of their production. At present, the microbiological industry of many countries produces a number of valuable exoglycans: dextran, xanthan (USA, France), pullulan (Japan), scleroglucan (USA), zanflo (USA), kurdlan (Japan). A number of other polysaccharides, studied in detail in laboratory conditions, tested in practice and produced on a semiindustrial scale, have been or are being introduced into production. The production of various polysaccharides is not universal. For each glycan, it has its own peculiarities determined by the physiology of the producer, the localisation and physicochemical properties of the polymer, and the field of its application.

The production of exopolysaccharides has advantages over the production of intracellular ones, as exoglycans are formed, as a rule, in much larger quantities and are more easily separated from biomass and purified from impurities. However, the production of exoglycans has its own technological difficulties. The accumulation of polysaccharide in the medium leads to restricted access of oxygen to the cells. In aerobic microorganisms this reduces the energy balance and inhibits polysaccharide synthesis. The increased viscosity of the medium makes it impossible to separate polysaccharide from the cells of the producer from the native culture fluid. It has to be diluted tens of times, and after removal of cells it has to be concentrated to the original or smaller volume. The solution to these problems is associated with additional costs.

Here are the main steps in the production of the most widely used common polysaccharides - dextran and xanthan.

Plasma substitutes from dextrans are produced under the names: clinical dextran, polyglucin, syncol, macrodex, plasmodex, chemodex, etc. To obtain dextrans, *Leuconostoc mesenteroid.es* strains are used. Fermentation is carried out on medium with 10-30% sucrose, dextran "inoculum", yeast extract, mineral salts. Conditions favouring the synthesis of the form of dextran used as a plasma substitute are created: linear glucan with more than 90 % of "-1,6-linkages, with a molecular mass of 60-80 thousand. For this purpose, the content of magnesium in the medium is limited, which stimulates the synthesis of branched dextran by introducing a "seed" in the form of dextran with a molecular mass of 20-30 thousand. Such an acceptor ensures the preferential formation of the required polymer.

The dextrans containing less than 70 % of a-1,6-glycosidic bonds, i.e. more branched ones, have the highest biological activity. Synthesis of biologically active dextrans promotes besides magnesium replacement of sucrose with molasses The optimum pH value for the growth of the producer lies within the range of 6.5-8.0, and for the accumulation of dextran-saccharase - about 7.0. Normally, the pH value of the medium is set between 7.0 and 8.0.

Bacteria break down sucrose into glucose and fructose. Fructose is digested by hetero-enzymatic lactic acid fermentation with the formation of lactic and acetic acids, mannitol and CO_2. Glucose polymerises into dextran. The process is rapid and the product can be isolated in 24h.

Dextran is extracted from the culture fluid, e.g. with methanol. Using certain techniques, it is possible to precipitate fractions of clinical dextran with a molecular weight of 60-80,000 even from a mixture of dextrans of different molecular weights. The entire product can be precipitated, dissolved in water and the required dextran isolated by fractionation. If necessary, the isolated dextran is depolymerised (enzymatically, by heat treatment or ultrasound). For purification the dextran is repeatedly dissolved in water, re-dissolved with me-tanol and fractionated.

Since dextrasaccharase is largely released into the medium and polymer synthesis takes place outside the cell, dextrans are also produced enzymatically. For this purpose, the producer is grown under conditions ensuring the highest activity of the extracellular enzyme complex. During the period of maximum activity of dextrans-charase, the culture liquid is separated from the cells and preserved by reducing the pH value to 5.0-5.2. At this acidity and at a temperature of about 15 °C, dextrasaccharase contained in the culture fluid remains active for at least one month. In the 80s of the XX century the technology of obtaining partially purified dextransaccharase was developed. The fermentation medium should contain sucrose and dextran "seed". The synthesis process lasts about 8 h. The enzymatic method is more convenient than the microbiological one, as it can be more reliably controlled and regulated, and allows to obtain dextran of the required molecular weight by varying the initial concentration of sucrose and enzyme, as well as the temperature of the process. This significantly simplifies and reduces the cost of subsequent technological operations. The use of immobilised dextransaccharases can be widely used in industry.

Xanthan produced by *Xanthomonas eanpestns* is available under the names: biopolymer Xs, keltsan, xanthan, keltrol. Bacteria are cultured on medium containing 1-5% carbohydrates (corn starch, raw sugar, molasses, etc.), organic nitrogen compound, double-substituted phosphoric acid potassium and trace elements, pH 6,5-7,2. Incubation, was carried out under aerobic conditions at 28 °C for 72 h. To improve polysaccharide properties, formaldehyde is added to the medium during fermentation. The addition allows to obtain glycan with increased resistance to various unfavourable factors, including temperature and salinity. The polymer is used as a solution of viscous culture liquid or as a powder dried in a stream of hot air. In the latter case, the polysaccharide is separated from cells by centrifugation and purified by precipitation with ethanol, methanol or acetone in the presence of an electrolyte.

Intensive searches for xanthan-type polysaccharide producers are conducted in various countries. The technology of domestic xanthan production on an industrial scale has been developed.

Often native polysaccharides do not have the desired qualities. They may be, for example, insufficiently active, or poorly soluble, or toxic, which prevents their use, etc. To improve the action of polysaccharide drugs to eliminate or reduce undesirable phenomena, i.e. to obtain a drug with the desired properties, isolated polysaccharides are sometimes subjected to chemical modification. For example, dextran is sulfated to give it anticoagulation activity. Treatment of native *A. faecalis* var. *myxogenes* glycan with solubilising agents allows to obtain derivatives forming helices without preliminary heating. The solubility of *A. puìlulans* glucan and *R. rubra* mannan can be increased by carboxymethylation. Reduction of toxicity of the antitumour drug protein-lipopolysaccharide complex from the culture fluid of *Serratia piscatorum* is achieved by its treatment with alkali. Obtaining polysaccharide derivatives is associated with additional technological operations.

The method of continuous cultivation of producers is becoming more and more widely used for the production of zkzoglycans. Many polysaccharides (xanthan, kurdlan, etc.), which are promising in practical terms, have already been obtained in a number of countries using this method. This method is very effective *and* economical, as it allows to obtain the product for a long time during its maximum accumulation and to use the substrate to the fullest extent. It is estimated that the percentage of conversion of carbon substrate into exopolysaccharides in flow cultivation is 2-3 times higher than in batch cultivation. Usually, nitrogen, as well as phosphorus and sulfur are used as limiting factors in chemostat. This ensures intensive accumulation of zooglycans. Due to proper selection of conditions, xanthan is obtained by continuous cultivation of the producer for many hundreds of hours.

The production of a number of other practically valuable polysaccharides is currently under development. A pilot plant at the Krasnoyarsk Medical Preparations Plant produces aubazidan and barium sulphate suspensions on aubazidan. A pilot plant for the production of mannan is being set up, and activities are being developed for the production of prodigisan and group-specific polysaccharides of meningococci (vaccines).

The microbiological production of polysaccharides faces a number of challenges, the most important of which is the replacement of expensive sugars with grasses.

In this regard, microorganisms are being sought to develop and form polysaccharides using hydrocarbons, ethanol, and methanol. In this connection, microorganisms that develop and form polysaccharides using hydrocarbons, ethanol, and methanol are being searched for. Alcohols as carbon sources are undoubtedly more convenient than hydrocarbons, as they are well soluble in water, which greatly simplifies the technology of isolation and purification of the product. For cultivation of microorganisms it is possible to use simpler media - synthetic or dialysis media, which are necessary for obtaining high-precision antigens. A cheap substrate, peat hydrolysate, is proposed to be used for the production of pullulan.

In order to obtain glycans with increased productivity, it is necessary to search for new polysaccharide producers with useful properties, especially extracellular ones, and selection and genetic studies of promising microorganisms. The method of experimental selection of active shtammers often turns out to be faster and more economical in comparison with their search in natural conditions. Thus, as a result of low temperature and subsequent irradiation with fast neutrons, it was possible to select *X. campestns* variants possessing increased synthesis of xanthan.

As micro-organisms are truly inexhaustible in polysaccharides, it is certain that among these polymers will be found substances which are chemically original and of great value to practice.

Chapter 15

OBTAINING ANTIBIOTICS

The principles of the production of motor metabolites are based on their formation by microbial cells. The biosynthesis of secondary metabolites is phase specific and occurs after the completion of the growth stage, in idioo phase, due to which they are called idiolytes.

Antibiotics are the largest class of pharmaceutical compounds synthesised by microbial cells. This class also includes antifungal agents, antitumour drugs and alkaloids. Globally, the annual production of antibiotics is $20 billion. US. Even more widely than in medicine, they are used in crop production, animal husbandry, veterinary medicine and food industry.

In 1928, the English microbiologist Flemnngwu established the ability of the green mould *Penicillum notanum* to kill microorganisms, and during World War II the industrial production of antibiotics was started. It should be noted that the therapeutic properties of moulds were described by the Russian dermatologist A. G. Polotebnov in 1871.

Antibiotics are low molecular weight effectors of originally natural origin capable of inhibiting the growth of living cells. Currently, 12,000 antibiotic drugs have been described, of which about 200 are used in the clinic. About 97% of known antibiotics are toxic, so they are not used in practice. Chemically, they are a composite group of organic substances

According to the type of action antibiotics are divided into bactericidal, causing the death of microorganisms, and bacteriostatic, disrupting the ability of microorganisms to divide. According to the spectrum of action, antibiotics of narrow and broad action are distinguished. The latter include tetracyclines, macrolides, amnloglycosides, prescribed in case of unidentifiable pathogens. They cause dysbacteriosis if used for a long time.

The peculiarity of antibiotics is the specificity of their action: they inhibit one or several processes only in some micro-organisms. Depending on this, the following groups of compounds are distinguished at the molecular level, which induce in bacteria:

. impaired biosynthesis of cell wall peptidoglycans,

- violation of individual stages of the translation process (translation is the realisation of genetic information);
- damage to the cytoplasmic membrane;
- disruption of nucleic acid biosynthesis;

. disruption of energy exchange.

Antibiotics have been widely used to decipher the mechanisms of protein biosynthesis, nucleic acids and bacterial cell wall structure.

The search for new antibiotics is driven both by the needs of practice and by the accumulation of resistant forms of microorganisms to many antibiotics. The enzyme lactamase, which converts penicillin into penicillinic acid, confers resistance to microorganisms. For this reason, new antibiotic analogues are produced using natural inhibitors of ß-lactamase, such as clavulanic acid and olivanoic acid.

Chemical methods of antibiotics production are very complicated and cannot compete with their biosynthesis by biotechnological methods.

There are several ways to produce both natural and synthetic antibiotics:

- fermentation of the microorganism-producing organism with a suitable precursor,.which induces nitetizantibnotics in the nondenaturing phase;
- use of "blocked" mutants for biosynthesis. Such mutants have blocked synthesis of the desired antibiotic. By exploiting the low substrate specificity of secondary metabolic enzymes and analogues of antibiotic precursors, they are converted into analogues of the antibiotic itself. This process is called biosynthesis or mutasynthesis:
- a sequence of reactions leading to the synthesis of an antibiotic is assumed:

$A \rightarrow B \xrightarrow{\text{фермент}} D \rightarrow E \rightarrow$ антибиотик; an antibiotic;

- lack of antibiotic synthesis in the "blocked" mutant:

$A \rightarrow B \xrightarrow{\text{фермент}} \underbrace{\boxed{C \rightarrow D}}_{\text{«блокированное» звено метаболизма}}$;

"blocked" metabolic link

- synthesis of the modified antibiotic after administration of the precursor analogue (D*):

$A \rightarrow B \xrightarrow{\text{фермент}} \ldots D^* \rightarrow E^*$ - a modified antibiotic.

Antibiotics are produced by mould fungi, actinomycetes, eubacteria and other microorganisms. Six genera of filamentous fungi produce about 1,000 different antibiotics, including penicillin and cephallosporium, and three genera of actinomycetes produce 3,000 antibiotics. Among actiniomycetes, the genus *Streptomyces* makes the greatest contribution, one of whose species, *S. griseus*, synthesises more than 50,000 antibiotics.

The process of cultivation of vdiolites consists of two phases (two-stage cultivation). The first phase involves the accumulation of a sufficient amount of biomass, which is grown on a suitable medium for microbial growth. This phase should be fast and the nutrient medium should be cheap. In the second phase, the start-up and active synthesis of the antibiotic takes place. Here the

fermentation is carried out on a productive medium. Antibiotic formation is regulated by the microbial cultivation conditions, so the optimisation of the nutrient medium is a major factor in increasing the yield of the product.
Most antibiotics are produced by deep aerobic batch fermentation under aseptic conditions.

15.1. Principles of antibiotic production

The widespread use of antibiotics in medicine, agriculture and other sectors of the national economy has set the task of producing these biologically active substances on a mass scale. This task was made possible by the creation of a powerful antibiotic industry.
The industrial production of antibiotics is based on a number of consecutive stages: obtaining highly productive strains-producers, development of the most favourable conditions for cultivation of antibiotic producers with maximum biosynthesis of this substance, selection and introduction into practice of appropriate methods of isolation and purification of antibiotics, creation of finished preparations and their quality control. Each of these stages should be supported by appropriate specialists (geneticists, microbiologists, technologists, etc.).
Today, antibiotic production is a powerful, well-developed branch of the pharmaceutical (in our country, medical and microbiological) industry. It occupies one of the leading positions in the production of pharmaceuticals. This industry is particularly well developed in the USA, England, Japan, France, Italy and other countries. For example, in the USA the annual output of antibiotics and their prodrugs amounts to hundreds of millions of dollars. By total production of antibiotics the CIS countries take the leading place in the world. According to Love and Elander, only 5-lactam antibiotics (penicillin and cephalosporin) account for half of the total volume of antibiotics produced.
The industrial method of antibiotic production is a complex, multi-stage process involving a number of technological stages. The main ones will be discussed below.
The so-called semi-synthetic method of antibiotic production is of great importance for obtaining various modifications of already known antibiotic substances with more valuable properties compared to the original preparations. This problem began to receive considerable attention since the early 1960s of the XX century.
The semi-synthetic method of producing antibiotics is based on the following principle. As a result of biosynthesis, initial antibiotics (penicillin, cephalosporin, tetracyclines, rifamycin) are produced and then chemically

modified. In a number of cases, not the entire antibiotic molecule but only its basic core serves as the main starting semi-finished product. For example, in the preparation of semi-synthetic penicillins and cephalosporins, 6-amino-penicillinic acid and 7-aminocephalosporanic and 7-aminodeza-cetoxycephalosporanic acids are used, respectively, which are then chemically modified.

6-Aminopenicillanic acid (b-APA) is obtained in three ways: as a result of the development of penicillin producer in a medium containing no precursor; chemical treatment (hydrolysis) of benzylpenicillin; enzymatic and enzymatic degradation of benzylpenicillinu

The possibility of obtaining 6-APC as a result of changing the conditions of cultivation of a penicillin producer was first pointed out by O. Kato in 1953. S. Betchel et al. in 1959 established that penicillin-forming strains of fungus are capable of forming 6-APC when developing in a medium containing no precursor. However, such obtaining of 6-aminopenncillanic acid has no practical, let alone industrial significance. This is due to the fact that, firstly, during the development of penicillin producer in media without a precursor, a relatively small amount of 6-APC is formed. Secondly, it is rather difficult to isolate this acid from the culture liquid and purify it from associated substances

A more efficient way to produce 6-APC is chemical, based on treatment of benzylpenicillin with phosphorus pentachloride to obtain an easily hydrolysable compound. This is then used to produce 6-APK. In this case, the yield of 6-APC is up to 95 %.

The enzymatic hydrolysis of penicillin molecule should be called the most rational method for the preparation of 6-aminopaicillic acid.

The enzyme penicillinacylase hydrolyses the benzylpenicillin molecule to form 6-APC and phenylacetic acid.

6-Aminopenicillanic acid is practically devoid of antibiotic activity.

Penicillinacylase can be produced by various types of microorganisms, including penicillin-producing fungi. The enzyme secreted by bacteria and proactinomycetes hydrolyse benzylpenicillin faster than other types of penicillins.

Currently, 6-aminopenicillanic acid is produced using immobilised penicillinacylase, which provides a high yield of the product and its high degree of purity.

The most important step after the fermentation of antibiotics is their isolation and purification from the culture liquid. In addition to the antibiotic substance, the culture liquid usually contains a large number of other impurities, often similar in their chemical and physicochemical properties to the antibiotic. The impurities accompanying the antibiotic substance are either substances of a mineral or organic nature, biosynthesis products, components of the nutrient medium or substances added to the culture liquid for its pretreatment. Their concentration often reaches several per cent and exceeds the concentration of antibiotics by tens and hundreds and sometimes thousands of times.

The isolation and purification of antibiotics is a complex technological process. The low stability of many antibiotics and the possibility of losing their activity during isolation and purification led to the predominant use of physicochemical methods of separation of substances, including sorption, extraction and crystallisation, i.e. such processes that were not accompanied by a sharp chemical impact on the antibiotic molecule.

Both equilibrium and kinetic methods are used for the separation, isolation and purification of antibiotics. However, equilibrium methods have proved to be more cost-effective and efficient for industrial applications. Here sorption and extraction methods have acquired the greatest importance. Equilibrium methods can be single-stage or multistage. In industrial applications and in the study of antibiotic properties, multistage processes are of great importance, as they can significantly improve the purity of the isolated drug.

Pretreatment and filtration of antibiotic culture fluids. Pre-treatment of the culture fluid and removal of the micellar mass is the first step before isolation and confinement. This is where the partial purification of the culture liquid from impurities begins. Depending on the properties of the antibiotic and the methods of its isolation and purification, the method of pretreatment of the culture fluid is chosen. The main objective of this process is to obtain a native solution (and in the case of antibiotics in mycelia - mycelial mass) with the highest degree of purity and the lowest losses, which allows for successful further operations of isolation of the antibiotic. Most antibiotics are isolated and purified from native solution by three methods: extraction, ion exchange and precipitation of insoluble compound. In the extraction method of extraction of antibiotics from liquid (penicillin, erythromycin, novobiocin) the native solution should be freed as much as possible from impurities capable of forming persistent emulsions with organic solvent. Protein impurities, which usually cause formation of persistent emulsions, are removed either together with the mycelium by various chemical methods, or by heating the liquid, or by both methods simultaneously. If the mycelial mass is easily removed without preliminary treatment, desemulsifiers are added to the active solution to keep the protein substances in a dissolved state under extraction conditions. In the extraction and purification of penicillin, Petrov's contact (paraffin contact, katexol, ul-travet, cetazol) is used as a demulsifier. In the extraction and purification of erythromycin, it is necessary to remove calcium ions from the native solution, which may contribute to the precipitation of the antibiotic during extraction.

In the case of ion-exchange method of antibiotic isolation from native solution, the latter should be freed from competing ions as much as possible in case of sorption of calcium, magnesium, iron ions on cationites. For calcium removal oxalic acid is used, for magnesium removal - phosphates, for iron removal - yellow blood salt. If the antibiotic is separated from the native solution by precipitation, it is desirable to remove all impurities from the native solution that can precipitate under these conditions.

One of the main tasks of culture fluid pretreatment is the coagulation of suspended particles. The importance of this task is evident in the coagulation and filtration of culture liquid of actiomncet or bacterial origin. Separation of the mycelial mass from the native solution in most cases is associated with considerable difficulties. This is due to the specific nature of the precipitate, which is usually amorphous, slimy, structureless and quickly clogs the pores of the filter material. The quality of the raw material and the raw material composition of the media have a great influence on the filtration process. For example, the use of soya flour, cake in the composition of the medium worsens the filtration of liquid. The use of hydrol instead of glucose as a source of carbohydrate reduces the filtration rate (streptomycin production). Incomplete consumption of nutrients and the use of fatty foams in the last stages of fermentation also impair filtration. In order to improve the filtration process, it is very important to stop fermentation in time - preferably at full carbohydrate intake, but before the onset of microbial cell destruction, as the filtration process of an autolysed culture is usually poor. In addition, increasing the duration of fermentation worsens the quality of the native solution, increases its pigmentation and the content of protein impurities. Antibiotic culture fluids are specially treated for coagulation. Depending on the properties of the antibiotic, the origin of the mycelial mass and the method of isolation and purification of the antibiotic, the culture liquid is treated to improve filterability: acid coagulation; introduction of electrolytes into the liquid; thermal coagulation; application of fillers; formation of filler directly in the liquid. Sometimes two methods are combined. Acid and thermal coagulation are used if the antibiotics are stable with changes in solution pH and temperature. Heating the liquid increases the filtration rate due to the coagulation and coagulation of proteins at high temperature and a significant reduction in the viscosity of the filtrate. Thermal coagulation increases pigmentation of the native solution and thus may impair the quality of the finished product.

Fillers such as diatomite or infuser earth, perlite, etc. are often taken to improve filtration. A good method of coagulating antibiotic culture fluids is to form the filler directly in the liquid by adding reagents that form an insoluble precipitate in the liquid. The precipitate in the liquid prevents mycelial particles from sticking together, promotes the formation of granules, so that the mycelium acquires a lumpy structure and forms permeable layers during filtration. The efficiency of this method is high and with the right choice of conditions can increase the filtration rate by 3-10 times.

The hydrodynamic conditions of the process have an extremely important influence on the coagulation efficiency of the culture liquid. Processing at low agitation allows to obtain well-filtered sediments in the form of small lumps. With increasing agitation coagulation becomes less distinct, the liquid settles slower. The precipitation after filtration is more moist and sticky. Hydrodynamic conditions during the processing of culture liquids have a great influence on their filterability. So, for example, exaggeration of the filtration rate of culture liquid from 1 to 2 m / s filterability of suspension and the degree of its dispersibility is almost the same. Further intensification of hydrodynamic regime (up to ω = 10 m/sec) leads to deterioration of filtration characteristics of suspension. Consequently, at filtration of culture liquids it is not the size of crystals, but the size of granules, which are formed at aggregation of particles of coagulant-filler and mycelium, that is of great importance. For filtration of the coagulated solution the duration of soak time after treatment is very important. Its reduction below a certain limit causes a decrease in the filtration rate and may in some cases lead to opalescence of the filtrate. The driving force of the filtration process is the pressure difference on both sides of the sludge layer, and one of its most important characteristics is the filtration rate, i.e. the amount of filtrate obtained from a unit of surface per unit time. The filtration rate depends on pressure, thickness of the sludge layer, structure and nature of the precipitated substance, viscosity of the liquid phase of the suspension and other factors.

15.2. Extraction processes

Extraction is widely used for the isolation and purification of many chemical compounds, including antibiotics (especially when they or the products of their compounds with carriers are not ionised in the aqueous phase). Despite the essential disadvantage of extraction processes, namely the use of harmful, explosive organic solvents, they are widely used in industry. Firstly, extraction processes are much faster than ion exchange processes in terms of time, distribution coefficients for some systems are very high, and it allows to reduce the volume of processed solutions. Ap- paraturally this process is very easy to realise in a continuous way. Extraction with a carrier is particularly interesting. Often this type of extraction is referred to as liquid and ionimobmeiom.

Extraction with a carrier. The transfer of the substance is carried out with the formation of a complex compound with a hydrophobic carrier not only at its expense, but also due to the suppression of polar groups in the antibiotic molecule by the carrier. Oleic acid, undecylenic acid, cetazol (cetylpyridinium bromide) are used as carriers. An acid (e.g. oleic acid) interacts with the streptomycin base to form a salt with a high solvation energy in the organic phase, and the substance is transferred to the organic phase. The carrier can be in either the aqueous or organic phase.

Penicillin is the most prominent representative of antibiotics, for which the extraction method for the isolation and purification of antibiotics is still successfully used. After pre-treatment of the culture liquid and separation of the precipitate, it is sent for extraction, which is carried out with butyl acetate at pH = 2; the distribution coefficient under these conditions can reach 30. Sulphuric acid is added to bring the pH of the extract to 2, thus suppressing the dissociation of penicillin in the aqueous phase, transforming it into undissociated penicillin acid, which easily passes into the organic phase. Then the butyl acetate extract is treated with a weak solution of alkali, and penicillin in the form of salt again passes into the aqueous phase.

The extraction is repeated again for more complete purification and concentration of the antibiotic.

Antibiotics, which are isolated and purified with the help of carriers, belong to the tetracyclic group (tetracycline, oxytetracycline). The presence of one basic and two acidic groups in their group allows the use of basic and acidic substances as carriers of these compounds As acidic carriers can be used sulfonic acids, phenols, fatty acids, and as liquid anionites - quaternary ammonium bases with a long hydrocarbon chain$_U$(C_{10}-$C_{(30))}$.

Schematically, the chemical reactions in this case can be represented as follows:

R - COO'Na$^+$ + oxytetracycline$^+$ θ RCOO - oxytetracycline$^+$ + Na ;$^+$

R - N$^+$ Hal$^-$ + oxytetracycline$^-$ θ RN oxytetracycline + Hal$^{(-)}$.

The studies conducted by *S.* I. Kaplan and other scientists show that the maximum degree of extraction (95-97%) is achieved when 5 % cetazol is used in a solvent taken in the amount of 15% of the volume of the native solution at pH of the aqueous phase of 9.5-10.2. A very important factor determining the process of antibiotic extraction is the duration of contact between the two liquid phases. The above studies have shown that at the optimum pH value, the equilibrium in the system is reached within the first minute of stirring. An important point is the subsequent re-extraction of oxytetracycline from the organic phase. The use of hydrochloric, phosphoric, citric acid for this purpose allows to obtain solutions with an antibiotic concentration of 40-50 thousand units/ml, but they have an intense dark colour. Re-extraction of oxntetracycline with 4-5% oxalic acid solution allows to obtain solutions with antibiotic concentration of 35-45 thousand units/ml with light colouring of the re-extract. Probably, this phenomenon can be explained by the different degree of dissociation of coloured impurities and the property of oxalic acid to interact with them.

Apparatus for extraction of antibiotics. Modern production of antibiotics is developing in conditions of increased introduction of a new progressive technological method of extraction (purification) - ion-exchange. In this connection in the apparatus of the extraction method the rational choice of design of apparatuses and mode of their work with application of economic indicators is especially actual.

In the chemical industry there are many extraction principles and types of apparatus suitable for the extraction of antibiotics. Whereas in earlier years the choice of extractors was made intuitively, nowadays, thanks to the work of Pratt, G. P. Pieterski andE. R. Valashek, D. E. Shkoropad and I. V. Lyskovtsev, the choice of extractors can be made rather strictly by rational evaluation of the advantages and disadvantages of each extractor.

The peculiarity of physicochemical properties of antibiotics - penicyl-lnn, tetracycline, erythromycin, bacitracin and their native solutions - allows to choose only extremely strict modes of their extraction.

Description of extractor design. The extraction process consists of two elements, carried out either in the same or in different apparates;

- emulsification of one liquid in another and their mutual movement, at which extraction takes place;

-Separations.

Modern continuous centrifugal liquid extractors are divided into direct-flow and counter-flow extractors according to the method of their operation. In direct-flow extractors, the solution and extractant, supplied by continuous streams, are mixed in a separate mixing apparatus or in the mixing device of the extractor-separator, then transported to the separator or separator drum of the extractor-separator. In countercurrent extractor-separators, the extractant

and the solution move countercurrently through the rotor apparatus.
In the case of separate continuous extraction and separation, straight-through mixer-extractors are used:

- 10-20 litre apparatus with stirrer or centrifugal pump;

. a tubular extractor, known as the Pietersky pipe, about 10 metres long, in which a developed turbulent liquid-movement regime is created;

. jet mixer of injector type.

The stirrer mixer is the least efficient because, unlike the other two extractors, the residence time of the liquid particles in the stirrer is different. In addition, the concentration of the antibiotic in the solution is equalised due to the operation of the stirrer. This reduces the driving force of the process (G. P. Pieterskikh and E. R. Valashek). For the separation of the resulting emulsions used plate separators of antibiotic liquids SAZh-ZM, equipped with twin centrifugal pumps with a head of 1 at.Productivity separators 2 5m/h. Separators *PSBS* (Germany), *De Laval* (Sweden), etc. are used abroad. At smaller production volumes supercentrifuges SGS-100 are used.

15.3. Sorption processes

Sorption methods for the isolation and purification of aitibiotics are the most widespread in industry. They have a number of advantages over other methods. The first sorption methods of isolation and purification of antibiotics were based on the use of molecular sorbents (activated carbon, aluminium oxide, etc.), which have universal sorption capacity, i.e. they sorb the isolated substance and a number of other impurities equally well. Great opportunities for the synthesis of ion-exchange sorbents, sorbents with different selectivity and especially with specificity towards certain antibiotics quickly brought them to the forefront.

Ion-exchange sorbents belong to the class of high molecular weight compounds, whose macromolecules have a net or spatial structure and in most cases are amorphous substances. The individual atoms of these giant molecules are connected to each other by covalent bonds. This structure of ion-exchange sorbents and the associated lack of solubility to a certain extent violate the generally accepted idea of electrolytes and ion-exchange processes between them. The dissociation of water-soluble acids or bases causes a change in the concentration of hydrogen ions. Ion-exchange sorbents also contain acid and basic groups, but their immersion in water does not cause a change in the concentration of hydrogen ions in it

Cationites are a special class of nknslot salts, which are characterised by multivalency and, having a bulky structure, are practically devoid of mobility. The cation included in the ionite is mobile, and only the force of

electrostatic attraction prevents its separation into the solvent, so an ionic atmosphere is created around the adsorbent. The cations in the ionic atmosphere are unequal in the strength of their electrostatic bonding with the adsorbent anion, so the process of ion exchange is a multistep reaction.
Hydrogen ions or hydroxyl groups of the ionite diffuse freely in the sorbent phase. All other part of the sorbent electrolyte, whose individual atoms and groups are connected by covalent bonds, is deprived of mobility, representing a giant anion (in case of cationite) or cation (in case of anionite) Thus, the process of ion exchange can be considered as an interaction of two electrolytes, one of which (sorbent) contains a practically immobile anionic (or cationic) complex. The process of ion exchange consists of diffusion of ions of the dissolved electrolyte to the surface of the sorbent grain, diffusion of ions of the dissolved electrolyte inside the sorbent, displacement of the mobile ion from the sorbent and diffusion of the displaced mobile ion from the sorbent-solution phase.
Classification of sorbents. Ionites, depending on the presence of ionogenic groups in them, can be divided into two classes: ion-exchange sorbents containing acidic groups in their structure - cationites (insoluble acids); ion-exchange sorbents containing basic groups in their structure - anionites (insoluble bases). The differentiation of ionites into acids and bases does not exclude the possibility of the existence of ampholyte ionites, whose ionogenic groups can behave as acids or bases depending on the pH of the medium. There are also ionites containing both acidic and basic groups at the same time. All materials used for sorption of antibiotics can be referred to one of the following classes: molecular sorbents, mineral ionites or no-exchange resins. Among the latter, carboxylic resins and sulphocationites, as well as anionites of different basicity are of the greatest importance.
The most common molecular sorbents used in the production of antibiotics are activated carbon and aluminium oxide.
Polymeric resin sorbents can be synthesised by polymerisation or polycondensation processes. The former usually have higher mechanical strength. However, a number of polymeric resin sorbents obtained by polycondensation are also successfully used in laboratory practice and industry.
Basic requirements for ionites. Usually the quality of ionites is reflected by a conditional assessment of the sorption properties of the ionite (total capacity of the sorbent, capacity of the sorbent at different pH of the medium, the rate of establishment of sorption equilibrium, the volume of desorbing solution and completeness of desorption); physical properties of the ionite in

certain conditions (swelling and strength of grains, heat resistance and chemical resistance, etc.). It is highly desirable that the ionite used for the extraction of valuable ions from solution or for the purification of a substance should have a maximum capacity. To fulfil this condition, it is necessary that the synthesized sorbent contains as large a number of ionogenic groups per unit weight as possible, fully ionised in these particular conditions of sorb cine. It should be remembered that ionisation of oxnfennl groups becomes noticeable at pH above 9, ionisation of carboxyl groups - at pH above 5, sulphogroups are fully ionised in the epe- de layer, amino- and imino-groups enter into ion exchange reaction in acid-loyereda, quaternary ammonium bases show properties of ion-exchange sorbent in neutral and even slightly alkaline media. Ionites used for chromatographic analysis should contain the same acidic or basic groups. In this case, it is easy to achieve a clear separation of the mixture of the corresponding antibiotics and other biologically active substances.

When selecting ionites, it is also necessary to take into account the relative strength of bonding of mobile ions with the ionite, so that their desorption is carried out as easily as sorption. The greater the ionic bonding energy between mobile and immobile ions of the sorbent, the more difficult it is to shift the equilibrium during the desorption period

The degree of swelling in sorbed solutions is of great importance for swelling ionites. With the increase of swelling increases the availability of ionogenic groups, which in turn increases the capacity of the sorbent and the rate of establishment of sorption equilibrium. The value of swelling capacity is closely related to the strength of sorbent grains. It decreases with increasing swelling. Under conditions of sharp changes in the degree of swelling, internal stresses arising in the grains cause their destruction. The mechanical strength of ionite D *is* defined as the ratio of the ionite volume after dust separation V_2 to the volume before shaking (%):

$$D = V_2/ V1 - 100.$$

Apparatus for ion exchange sorption of antibiotics. The apparatus for ion exchange sorption processes is undoubtedly simpler compared to extraction processes and is affordable.

The design of an ion exchange filter, also called a column, for sorption under dynamic conditions must ensure:

. the constant presence of resin underneath the solution bed;

- minimal carry-over of fine resin grains with the escaping filtrate;
- minimal traceability of the resin;
- the absence of dead spaces in the erbent;

- insignificant dilution of the treated solutions with water.

The design, size of the colony hydrodynamic mode of operation depend on the mechanical properties of the resin, its sorption capacity and cost. Ionites used for water softening and desalination, e.g. sulphogly, are usually solid particles, resistant to brushing, creating low resistance to liquid flow in the nozzle. The high strength of resin allows to apply as filters the usual capacitive apparatuses with elliptical lids and with support devices, distributors and siphons. Such filters due to small capacity of resin and high capacity have diameter and height of resin layer about 1,5 m. Water is fed into these filters from the top, but from the bottom - only for periodic loosening of the ionite layer after regeneration.

For demineralisation of streptomycin eluate and other antibiotics, ion-exchange filters of similar design are used, although of smaller size, because inexpensive resins used for this purpose, such as SBS-1 and EDE-1O, have mechanical properties of high level.

For sorption of antibiotics more expensive ion-exchange resins are used, e.g. cationites KB-4P-2, KB-2 of weakly cross-linked large-porous ion-exchange resins with relatively low mechanical strength. In the swollen state, being in a thick layer, they are deformed under the action of gravity and pressure of the downward flow of solution. In this case the porosity of the nozzle becomes much less than the porosity of spherical non-deformable particles (equal to 0.4), as a result of which the hydraulic resistance to the solution flow increases and the area of contact of resin grains with the solution decreases. In this regard, domestic enterprises have switched to the use of upward flow of solution. The solution is fed at a slightly higher critical velocity (related to the full cross-section of the apparatus), at which the resistance of the layer becomes equal to the weight of the layer and the grains become suspended. Sorption in such a fluidised bed proceeds faster, the sorption capacity of the resin and the purity of the eluates increases.

For a polydisperse layer, the equivalent ze- ria diameter is calculated. The critical velocity is in the order of tenths of mm/sec. In order to prevent carry away of fine resin particles, a filter design in the form of a 0.4-0.7 m diameter column with an extended open top part made of organic glass is used. In the upper part of the column the velocity of the solution is reduced to values that are much lower than the critical velocity. The resin carried away by the flow is captured in troughs and returned to the working part of the column.

The upward flow is also used in regeneration, where the conversion of resin to another form, for example from the hydrogen form of cationite to the sodium form, is accompanied by swelling of the resin and approximately doubling of its volume. If the solution is fed from above, this resin swelling can destroy the column. The upward flow is also useful for washing and loosening the resin bed. Antibiotic desorption, accompanied by resin compression, is performed with a downward flow to increase the eluate concentration.

Any ionic filter has a drainage or support layer, a lower and upper distributor and a device (siphon) to maintain the solution level 0.1-0.15 m above the resin level. The drainage layer is used to maintain the resin layer. For filters with a diameter of more than 1.5 mm it consists of a layer of gravel, anthracite or other inert material. At smaller diameter of filters porous plates or conical caps made of plastic with slits having a cross-section equal to the fine fraction of resin grains, which are strengthened in a holey false bottom, are used. Filter fabric is used less frequently. Such a drainage layer also serves as a distributor. The distributor in the form of a holey pipe, installed directly above the resin layer (taking into account the swelling), serves ^ uniform distribution of the liquid flow across the cross-section of the filter. Uneven distribution strongly worsens the efficiency of the sorption process and leads to the appearance of dead zones, which are not washed by the solution.

Ion-exchange filters are usually made of carbon steel. To protect against corrosion caused by acid and alkali solutions, gumming or glass lining of the apparatus body and pipelines is used. This protection is effective due to the absence of organic solvents and high temperature solutions. Columns with a diameter of less than 0.3 m should be made of vinyl plastic, organic or ordinary glass. Transparent bodies of columns are very convenient in work. Vinyl plastic pipework is not strong enough, so it should be avoided during prolonged operation.

Ion-exchange apparatus and technology are developing on the way of acceleration of sorption process, automation of preparation of solutions for regeneration, automation of control of concentration of solution flowing from the column, transfer of ion exchange process to continuous one.

15.4 Crystallisation

The crystallisation process in the production of antibiotics is usually the final stage and therefore requires careful research.

The overall quality of the product depends on the way in which it is isolated and chemically purified from the culture liquid and, in particular, on the correct adherence to the crystallisation conditions at the final stage.

Nowadays, the requirements to the quality of manufactured drugs are increasing, and this makes it necessary to delve even deeper into the processes of crystallisation and search for ways and possibilities of their use for obtaining drugs in finished form. Solids can be crystalline and amorphous. The crystalline state is distinguished from the amorphous state by the arrangement of molecules, atoms or ions in a definite and strict order. X-ray analysis has shown that many substances once thought to be amorphous have a regular arrangement of molecules, but the term "crystalline" is most often used to denote a high degree of internal ordering resulting in the formation of certain outer faces of a crystal. In gases and fluids, the movement of molecules is free and disordered, so the physical properties of these substances are the same in all directions, while most crystals have different mechanical, electrical, and magnetic properties in different directions.

Different substances that have almost identical crystalline forms when crystallised are called isomorphic. They are often similar in chemical composition and have the same chemical formulae. Substances capable of crystallising into different but chemically identical forms are called polymorphic.

The basic conditions of the crystallisation process are to obtain a supersaturated solution; a solution in equilibrium with a solid phase is said to be saturated with that solid phase. From a saturated solution, it is relatively easy to obtain a solution containing a higher percentage of dissolved solid phase. Such a solution is called a supersaturated solution.

If the solution is cooled without loss of solvent, spontaneous crystallisation will not occur until the necessary conditions are reached. At this point crystallisation may proceed spontaneously, or it may be induced by inoculum, agitation or mechanical shock. Further cooling to some new point may be necessary to initiate crystallisation, especially if the substance has good solubility. Although the tendency to crystallise increases once the labile zone has been passed, sometimes crystallisation does not occur due to an increase in the viscosity of the solution and the transition to a glassy state. The supersaturation of the solution can be achieved by removing a certain amount of solvent from it by nsparenne. The transition through the supersaturation curve into the labile zone is rarely carried out, since the surface from which evaporation proceeds is usually supersaturated to a greater extent than the bulk of the solution. Crystals formed on this surface eventually enter the solution and etch it before the desired conditions are reached in the bulk of the solution.

In practice, a combination of cooling and evaporation is most often used.

Crystallisation and dissolution. Usually the processes of crystallisation and dissolution are considered equal and interrelated, and if they were diffusive in nature, the rate of crystallisation would be equal to the rate of dissolution at a given temperature and with the same driving forces, i.e. with the same deviations from the equilibrium saturated states; all faces of the crystal should grow and dissolve at the same rate. Such conditions are rarely achieved in practice. Crystals usually dissolve faster than they grow, and different faces usually grow or dissolve at different rates, although this is not always the case. Some researchers believe that crystals dissolve faster because the exposed solid surfaces are not the same in each case. When a crystal grows, its faces are flat, but when it dissolves, its faces are usually pitted, which leads to an increase in the area of contact between the solid and liquid phases. The solubility of crystals depends on their size. The solubility of very small crystals increases considerably, and if the crystal size is too small, small crystals will disappear from the system and large crystals will grow. This phenomenon causes some difficulties in the initial phase of crystallisation. The first crystals (germs) cannot form because of their very small size and high solubility. The size of crystals depends on the surface tension of the solution, which can vary depending on the content of impurities present. The greater the surface tension, the stronger its effect in the direction of reducing the total interfacial surface at the solid-liquid boundary. This condition leads to the formation of larger crystals, because their specific surface area (per unit weight) is smaller than the specific surface area of small crystals.

. It should be noted that large crystals cannot have a high degree of purity. Due to agglomeration they contain inclusions of mother liquor, while small crystals are characterised by homogeneity, but their large surface area makes them difficult to wash.

Crystal swarming. As soon as stable nuclei, i.e. particles larger than a critical size, have formed in a supersaturated or supercooled system, they begin to grow, becoming crystals of visible size. The growth of crystals depends on very many reasons. Let us dwell on some of them.

A whole series of curves for many systems was obtained by D. Tamman. For all the systems studied it was found that the optimum temperature for the nucleation of crystallisation centres was much lower than that required for maximum crystal growth. At temperatures below the optimum temperature the viscosity increased to a value that prevented crystallisation, while above the optimum temperature the nucleation of crystallisation centres was hindered by the intense molecular motion of the liquid.

Influence of stirring. The rate of crystal growth, at a given temperature under constant supersaturation conditions, can be significantly altered by agitation of the liquid or by rotation of the crystal relative to the liquid. The rate of growth increases considerably in the first stages, but soon conditions are reached where further agitation has no effect. From a number of similar studies, it has been. concluded that diffusion is not the only and most important factor to be considered in the study of the process of crystallisation. It should also be noted that in a crystalliser in which the crystals are kept in suspension by agitation, large crystals will grow more rapidly than small ones, as the former will be favoured by the higher relative rates of reaction between the solid and the solution.

Shape.of.crystals. The shape.of.crystals depends on very many factors: e.g. solvent properties, pH of the solution, presence of impurities, degree of supersaturation or supercooling, cooling rate, crystallisation temperature, intensity of stirring, etc. Rapid cooling of the solution often causes preferential crystal growth in one direction, resulting in the formation of needles.

The most frequent case is probably the change of the crystal shape in the presence of impurities in the crystallising solution. Shape change under the influence of impurities is essentially a surface phenomenon: impurity molecules or ions are attracted to the various faces of the crystal and are physically or chemically absorbed at the surface. This phenomenon reduces the area suitable for the nucleation of crystallisation centres on the surface for the deposition of solute, and growth on that face is retarded.

Crystallisation of tetracyclines. The presence in the molecule of tetracyclines of two acidic non-dione axisovion groups makes it possible to obtain them in the form of a base or a salt of an acid, and the ability to complex with heavy metals makes it possible to precipitate them even from native solutions in the form of calcium salts. The solubility of these compounds in water is different and depends on the pH of the aqueous phase. Tetracycline and oxytetracycline have the lowest solubility at pH values from 4.0 to 7.0. This property of antibiotics of the tetracycline group allows their precipitation from aqueous solutions in the form of insoluble crystalline compounds. For example, the precipitation of oxytetracycline base is carried out as follows. The extract from the previous stage is treated with activated carbon in an amount of 1 % to remove residual extractant and pigment substances. The charcoal is separated and a 10% sodium hydroxide solution is added to the clear filtrate under stirring until a stable pH value of the aqueous phase of 4.2-4.5 is obtained. The precipitated crystalline precipitate of the oxantetracycline base is filtered from the mother liquor, washed with water and dried in vacuum at 50 degrees centigrade.

High hydrophilicity of oxytetracycline chlorohydrate prevents its isolation from aqueous solutions even at high concentrations. For this reason, to obtain crystalline oxytetracycline chlorohydrate it is necessary to use an organic solvent, in particular methyl alcohol.

According to the data obtained, the solubility of oxytetracycline base in methanol at 20 ^{O}C is 7.5 thousand units/ml. It is also known that the solubility of oxytetracycline base in methanol increases in the presence of calcium chloride. Probably, this increase in the solubility of oxytetracycline base in the presence of calcium chloride can be attributed to the ability of tetracycline bases to form complex compounds with heavy metals (like chelates).

It was noted that the solubility of oxytetracycline base in methyl alcohol increases strongly with increasing concentration of calcium chloride. . However, as the concentration of oxytetracycline increases, the viscosity of the solution increases, due to which its filtration rate decreases and crystallisation conditions deteriorate. On the other hand, the solubility of oxytetracycline hydrochloride in acidic methanol decreases with increasing calcium chloride concentration. Comparison of these data allowed us to conclude that the optimum condition for obtaining crystalline oxytetracycline hydrochloride is a base concentration of 100-120 thousand units/ml in a 7-9 % solution of calcium chloride in methanol. When acidification of such solution with hydrochloric acid occurs rapid crystallisation of hydrochloride, the solubility of which in methanol solution of this composition is minimal.

The crystallisation process is carried out as follows. Technical base of oxytetracycline is dissolved in 8 % methanol solution of calcium chloride to obtain the above concentration. Concentrated hydrochloric acid is added to the solution treated with activated clay at 3...5 ^{O}C. The mixture is stirred for 30 min at the same temperature, after which the crystalline precipitate of oxytetracycline hydrochloride is filtered off and washed with cooled methanol. The precipitate is dried in vacuum at 35...38 °C to obtain oxytetracycline hydrochloride in the form of lemon-yellow crystals with an activity of 900 U/mg. The crystallisation process of two representative antibiotics of the tetracycline group (tetracycline, chlortetracycline) proceeds and is carried out in a similar way, with minor details and changes in each case.

It should also be pointed out that the crystallisation process both "in the case of obtaining the base and in the case of obtaining oxytetracycline hydrochloride is crystallisation with chemical reaction.

Crystallisation of penicillin. Exceptionally high hydrophilicity of potassium and sodium salts of penicillin does not allow their crystallisation from aqueous solutions. Crystallisation of potassium salt of penicillin is carried out from butyl acetate extract with addition of saturated aqueous solution of potassium acetate. In this case penicillin, being in butyl acetate in the form of penicillinic acid, reacts with potassium acetate. As a result of chemical reaction the potassium salt is formed, which due to small solubility in butyl acetate precipitates. In this case crystallisation proceeds with chemical reaction. The precipitated crystals are washed with butanol to remove potassium acetate and dried. The sodium salt of penicillin is crystallised by azeotropic distillation of the solvent. Butanol is added to the aqueous solution of penicillin in a vacuum evaporator, which with water forms an azeotropic mixture consisting of 68 parts of butanol and 32 parts of water. The azeotropic mixture of butanol and water is distilled off at 20 ^{O}C and residual pressure 10 mm Hg. On complete removal of water from the cube residue and cooling, the Na-salt of penicillin crystallises, which is then also washed with butanol and dried.

15.5 Drying of antibiotics

Due to the instability of antibiotics even in a slightly moistened state, it is necessary to dry the preparations to a residual moisture content of 0.5-2.0 %. Depending on the aggregate state of the antibiotic at the end of purification and its stability at higher temperatures, four drying methods are used.

1. Antibiotics precipitated in the form of crystals, which have a relatively high stability (e.g. tetracycline) are dried at atmospheric pressure and temperature up to 90 ^{O}C in ka- merihn or pneumatic dryers.

2. Precipitated low-stable antibiotics (penicillins) are dried in vacuum dryers at a technical vacuum and a temperature of about 40 C in cabinets, dryers-venulet ndr. Antibiotics obtained during isolation-purification in the form of a concentrate, i.e. 5-15 % aqueous solution (streptomycin, antibiotics of the neomycin group) are dried by two methods to exclude inadvertence. Antibiotics obtained during isolation-purification in the form of concentrate, i.e. 5-15 % aqueous solution (streptomycin, antibiotics of neomycin group), which are very unstable in the dissolved state, are dried by two methods excluding inactivation and deterioration of the quality of preparations.

3. Slow drying (several hours) is carried out at sub-zero temperature by sublimation of water from a frozen solution under medium or deep vacuum. This is called molecular drying.

4. Rapid drying (within fractions of a second) takes place at high temperatures of about 130 ^{O}C in the form of an aerosol formed from the solution in a current of hot air. This drying is called spray drying.

Molecular drying. The method of sublimation drying, first discovered and patented in 1921 by the Soviet engineer G. I. Lappa-Starzhenetsky, consists in the fact that a wet material or solution (bacterial mass, blood serum, antibiotic solution, fruit juice, fruit, meat, fish), frozen to a temperature of -20...-40 °C, is dried in vacuum with a residual pressure of about 0.01 mmHg by distillation of water from crystals.

In this case, due to vacuum, the average free path length of water molecules in the pores of the dried material (~5 mm) is much longer than the cross section of capillary pores (KUMo;^{2} mm) of the frozen material, through which water vapours are removed. Under these conditions, water molecules move in the pores in the form of molecular bundles (effusion), so the freeze drying method is called molecular drying at the suggestion of A. V. Lykov.

The technology of such drying is as follows. In order to prevent its foaming in the freezing chamber and its release into the vacuum system, the biotic solution is pre-freezed in the chambers of conventional refrigeration plants to -40 ^{O}C temperature. The solution is sterilised by filtration before freezing and poured into vials using a semi-automatic dispenser, 3-5 ml depending on the clinical dose of the antibiotic. The faster the freezing process, the more finely crystalline and more homogeneous the structure of the ice formed by the antibiotic concentration and the faster the subsequent drying due to the more developed sublimation surface.

Moisture is removed at a constant rate during sublimation while maintaining the shape of the frozen material or solution. Since drying is carried out from the surface of the body, the finer the dosage, the faster the drying is completed. Its optimum mode corresponds to the temperature and vacuum at which the sublimation rate is the highest. This is achieved by drying at a constant temperature slightly below the melting point of the product - the cryohydrate point. Melting of the product is inadmissible as foaming of the solution causes its ejection from the bottle or partial transfer to the walls of the bottle. Such "foamed bottles" are discarded (n)pH"TMe quality) TM,(Y1)_ in pePTMD sublimation (5-8 ^ removes about 80% of moisture at a temperature of frozen product-35 ... -40 C and residual pressure of 10-30 mm Hg. At the end of the sublimation period, the temperature of the inner layers of the material rises to 0 °C and the period of evaporation of residual moisture begins. At this stage, the material is gradually heated to ambient temperature (temperature of the heat transfer medium) and the drying rate gradually decreases to zero.

Sterile antibiotic solutions are dried in open vials, placed in cassettes and covered with a cloth. The cassettes are placed on several boats of a rectangular or cylindrical cabinet-sublimator. In the hollow shelves of the cabinet after creating a vacuum in the system, first cold and then heated water is fed to supply the heat required for sublimation. At the end of drying, the water temperature is raised to 80 ^{O}C. The vacuum in the cabinet is created by two stages of pumps. Forvacuum (or gas-ballast) by design rotary-oil pump, switched on at the beginning of drying, creates a residual pressure of a few millimetres of mercury column. The vapour-jet diffusion pump then reduces the residual pressure to 10-50 μm. To reduce the load on the pump and to protect it from condensate water vapour, a scraper condenser or, in the case of large plant capacities, a tube condenser is provided in the scheme. In the scraper condenser, water vapour condenses and freezes on a cylindrical surface cooled with refrigerant to temperatures lower than in the sublimator, forming a layer of ice. The ice is continuously scraped off with a scraper into the ice receiver, from where it is periodically removed.

Spray drying. Spray drying is one of the most modern and promising methods of dehydration of medicinal solutions of thermolabile nature and food products (milk, eggs). The principle of this method is that the solution to be dried is atomised with the help of nozzles, jet air or a rapidly rotating disc to particles of 5-25 microns in the current of air flowing through the drying chamber, heated to a temperature of about 160 ^{O}C. The surface area of the particles is about 0 0.5 cm. A particle surface area of approx. 0.5 m^2 per 1 m^3 of solution ensures drying within a fraction of a second. The dried product in powder form is separated from the waste heat transfer medium in special dust separators.

Film footage of the drying process of single drops of streptomycin solution, conducted by O. A. Kremnev et al. allowed us to identify five stages of drying drops of colloidal solutions: drop heating; drying at a constant rate at a temperature equal to the temperature of a wet thermometer (about 40 °C); formation of a crust of dry streptomycin on the drop surface at a humidity of 20.0-25.0 %, preventing the escape of vapour, and raising the drop temperature to the boiling point (-105 °C); drying of the drop at a constant rate at the boiling point; drying at a falling rate approaching the ambient temperature.

This character of the droplet drying kinetics allowed us to conclude that it is expedient to divide the process of dehydration of highhmoisture antibiotic solutions into two stages in order to intensify and cheapen the process.

1. Since at the stage of evaporation from the initial humidity of 16.0-20.0 % to the crusting humidity the drop has the lowest temperature, the duration of its stay in the chamber can be chosen arbitrarily. It is advisable to use the heat-carrier to the maximum extent, i.e. remove it with a temperature 3-5 ^{O}C

above the solution temperature, and bring the relative humidity to 90 %. The speed of the heat-carrier through the chamber to intensify the process can be taken in 3-8 times more than in the drying chamber, and the size of the chamber to make 1.5-2 times smaller than the size of the torch of atomised material.

2. Drying from the moisture content of coconut formation (20.0-25.0 %) to a final moisture content of 1.5 % is carried out within a minimum time, since evaporation of the thermolabile solution takes place at its boiling temperature. This is achieved by using very dry air (final relative humidity 4-6 %) and low consumption of highly concentrated solution. Such conditions also make it possible to increase the speed of the coolant compared to a single-stage dryer. The spent coolant from the drying stage at a temperature of 100...105 ^{O}C is fed to the evaporation stage and is fully utilised.

Application of such two-stage evaporative-drying method of streptomycin dehydration at the Kiev plant of medical preparations allowed in comparison with one-stage spray dryer of the Danish company *Niro* to increase the coefficient of heat utilisation of the coolant from 25 to 75 %, increase the yield of antibiotic (taking into account losses on vacuum-vaporisation, necessary at one-stage mode) from 85 to 93 % and reduce the cost price of the process of streptomycin dehydration from 0,98 to 0,34 p/kg streptomycin.

The air is sucked from the atmosphere through a fabric filter of coarse cleaning by a fan and blown through a Petryanov filter, a steam heater and electric calorifiers into the evaporator stage of 7 m capacity. Air with a temperature of 165 ^{O}C is supplied to the chamber through a gas distribution device. Sterile streptomycin solution is fed to a disc rotating at a speed of 18 thousand rpm. A stream of air passing between the ceiling of the chamber and the radial plume of atomised solution prevents the solution from settling on the ceiling of the chamber Then the air turns downwards. The evaporation of the solution reduces its volume by a factor of four. The mortar with a humidity of 25.0-30.0 % is separated 98-99 % from the air in a wet cyclone. The air is discharged into the atmosphere and the liquor is immediately sprayed on the drying chamber disc with a capacity of 17 m^3. The heat transfer medium is supplied to the drying chamber from a similar system of atmospheric air sternlisation and heating. The antibiotic powder is separated from the exhaust air in a cyclone and discharged into a hermetically sealed device for sterile packing of the antibiotic into containers. The exhaust air containing approx. 7 % of the antibiotic powder is directed to the evaporation chamber, where it is used as a secondary drying agent due to its low humidity.

The drying chambers are made of polished stainless steel and have the shape of a cylinder with a cone bottom.

The dryer has a capacity of 200 l/h of evaporated moisture. A system of automatic devices regulates the temperature of the heat carrier in the inlet chamber and the solution flow.

Comparison of molecular and spray drying. From an economic point of view, the two methods are currently roughly equivalent. Molecular drying requires 2.7 times more production area and 2.1 times more energy consumption, as well as higher labour costs. At the same time, the yield of antibiotics in molecular drying (97-98 %) is higher than in spray drying (92-94 %). If we take into account the losses during packing of antibiotic powder, the decrease in yield below 90%, observed in the practical work of enterprises, nullifies the economic advantages of the atomising method. With a more thorough development of the mode of the recently developed two-stage spray dryer and elimination of design deficiencies, its economic performance will undoubtedly be higher than that of molecular drying.

However, molecular drying has advantages in the quality of preparations - a more important indicator for antibiotics for medical parenteral use than the economic one. Drying at negative temperature under vacuum ensures preservation of quality - solubility, colourlessness, apirogenicity, absence of opalescence of solutions, etc. There is no need for subsequent packaging, and this ensures the preservation of sterility

In spray drying, 7% of the antibiotic returned from the second stage to the first stage is re-dried. This and some other technical features of the process (e.g. polydispersity of the spray) reduce the quality of the product. As the spray drying process is adopted, the quality of the dry product will improve.

In the production of antibiotics, both methods considered for drying antibiotic solutions are promising.

15.6. Antibiotic use

Antibiotic substances are widely used in medicine, agriculture, food industry, as specific inhibitors in biological research.

Antibiotics mediation. The discovery of antibiotics has revolutionised medicine. Many antibiotics have proved to be indispensable therapeutic agents in the treatment of infectious diseases that were previously considered incurable or highly fatal. These include some forms of tuberculosis, especially meningitic tuberculosis, which before the use of antibiotics was 100 per cent fatal, plague, Asian cholera, typhoid fever, brucellosis, pneumonia, various septic processes andr.

Some antibiotics can inhibit the development of malignant tumours and are active against a number of viruses.

To date, about a hundred antibiotics have been used in medical practice. The search for new antibiotic substances, obtaining valuable semi-synthetic preparations of antibiotics expands the possibility of their practical application in medicine.

Antibiotics in agriculture. First of all, in agriculture, antibiotics are used as preparations in veterinary medicine for the treatment of various diseases in farm animals. In this case, as in medicine, they have proved to be very effective agents. Antibiotic substances are also used in the fight against phytopathogenic organisms - plant pathogens that cause considerable damage to agricultural production. When selecting an antibiotic used in crop production, the following basic requirements for the pre[parate;] are guided,

- specific biological activity against plant pathogens;
- the ability to easily penetrate plant tissues and exhibit intrinsic biological activity;

. the harmlessness of therapeutic doses to the plant,

. when found on the surface and inside tissues.plant relatively long time of biological activity, and also easy and rapid inactivation when ingested into the soil.

One of the most important requirements for antibiotics used in agriculture should be the fact that these preparations are not used in medical practice, which is a fundamental way of reducing the emergence of resistant forms of mycro-organisms that are pathogenic to humans.

Different antibiotics can be used to control phytopathogenic organisms.

Griseofulvin is produced by mould fungi of the genus *Retsglit (R. iirticae, R. mgncans, R. rustnchi*). It is used against phytopathogenic fungi, primarily fungi belonging to the genus *Botrytis.* It is active against the causative agent of rust, powdery mildew, cabbage boil.

Trichothecine is produced by mould fungus *Trichothecium roseurn* and inhibits the development of a number of phytopathogenic fungi, including *Botrytis cinerea, Helmintosporium.* Once in the soil, trichothecin is inactivated by the enzyme trichothecinase formed by soil fungi of the genus *Fusarium, Aspergillus, Penieillum.*

Kasugamycin, produced by *Streptomyces kasugaensis*, is active against the fungal disease of rice Pnrnculariosis, which is widespread in Japan. The disease is caused by the fungus *Pelicularia oryzae.*

Polyoxins are antibiotics belonging to peptidyl-pyrimidine-nucleoside compounds with antifungal activity. They are produced by *Streptomyces cacaoi* culture *and* inhibit the growth of phytopathogenic fungi of *Alternaria,*

Cochliobalus, Pirulana genera.

Validamycin A, produced by *Streptomyces hygroscopicus* var. *limoneus,* has biological activity against the phytopathogenic fungus *Rizoctonia solam, the* causative agent of rice disease. The total formula of validamycin A is known: $C2_0H_{35}NO_{13}$ - H_2O. The antibiotic is easily degraded by soil microorganisms.

Tetranastine is an antibiotic substance of the actinomycete origin. Its producer is *Streptomyces aureus*. It has specific activity against parasitic spiders and mites of fruit-trees.

Herbicidins A and B, synthesised by *Streptomyces saga-,* inhibit the development of the rice pathogen *Xanthomonas oryzae.* Herbicide A delays germination of rice and Chinese cabbage seeds and has selective herbicidal activity against dicotyledonous plants. By chemical structure and biological activity herbicidins A and B are close to Toyocamycin formed by *Str.toyocaensis* culture. _

Antibiotics in the food industry. In addition to physical and chemical methods, antibiotics are also used to control microorganisms that cause food spoilage. However, antibiotics used in medicine cannot be used for these purposes. This rule was introduced in our country and other countries and is related to the prevention of the emergence and spread of antibiotic-resistant microbial moulds.

Antibiotics used in the food industry include subtylin and some others.

Subtilinobraised by *BaaUm ..bish* culture and is a polypeptide. It is active against gram-positive and gram-negative microorganisms, including acid-resistant bacilli. By using subtilin in canning vegetables, it is possible to significantly soften their heat treatment, which is of great importance for the preservation of vitamins, flavour and consistency of the product.

Nisin is a high molecular weight peptide, or perhaps even a low molecular weight protein formed by *StrepPKOccuslact**

Nisin is not used in medical practice. It is used in canning tomatoes, green peas, cauliflower and other products. Good results are obtained in preservation of cheeses.

The antibiotic inhibits the development of a number of thermophilic spore-forming bacteria, is not toxic to humans.

The use of antibiotics in the food industry should be under constant and strict control of specialists and the relevant competent authorities.

Chapter 16

MELANIN PRODUCTION

Melanins (Greek: *rnelanos* - black) are usually black or dark brown pngmeityjnvotihy, plant imimicro-organisms. In higher animals and humans, melanins are the main group of pigments. In animals, melanins give colour to wool, in birds - to plumage, in humans they are responsible for the colour of eyes and skin.

Melanins are amorphous high molecular weight substances. Melanin pigments, present in all living organisms, constitute an extensive group of natural condensed phenols. They can be characterised as a class of compounds with low toxicity and diverse biological effects. These properties may find wide application in preventive and clinical medicine. Melanins of mammalian" plants are more deeply studied: they are not soluble in water, mineral acids, organic solvents; they are well soluble in alkalis and then precipitate when solutions are acidified, which is used for their isolation. Melanins are chemically destroyed by heating above 200 °C, fusion with alkali, oxidation with concentrated solutions of $KMnO_4$nlyH_2O,

According to the precursors of melanins in the organism, they are divided into eumelanins, pheomelainins and allomelanins. Eumelanins (black) and pheomelanins (yellow, red and brown) are common in animals, while allomelanins (black) are found in plants, fungi and bacteria. The precursor of eumelanins is tyrosine, from which pigments containing C, H, N and O are formed in the body. The precursors of phe- omelanins, tyrosine and cysteine, are transformed in the body into sulphur-containing pigments. The precursors of allomelanins are diphenols (pyrocatechin, etc.), from which nitrogen-free melanins are formed. In organisms, melanins are in complex with proteins.

Many of the properties and biological functions of melanins are determined by their ability to function in the body as a phenol-semiquinone-quinone system. The mechanism of melanin formation has not been fully elucidated. It is believed that the first stages of melanin biosynthesis are enzymatic and catalysed by oxidoreductases (e.g., tyrosinases, polyphenol oxidases, etc.), and the subsequent stages occur spontaneously with the participation of free radicals.

Melanins can be obtained by autoxidation of 3,4-dihydroxy-phenylalanine (DOPA), 5,6-dihydroxyindole and other diphenols.

Melanins are characterised by the presence of unpaired electrons in their structure and have the properties of stable free radicals. This feature of melanins is important for their protective function in the body. They not only absorb various radiations, but also neutralise and neutralise free radicals dangerous for cells, which are formed during the action of ionising radiation and some chemical substances on living organisms.

Melanins can exist in several redox states, have electron-transport properties,[(s)],[(mi)]-

In animals and humans, the process of melanin formation in the body takes place in special cells - melanocytes - and is regulated by the hormonal system, mainly with the help of hypophysis hormones ("- and ß-melanocyte-stimulating hormones). Some physical factors (sunlight, UV and X-rays) and chemical reagents stimulate the formation of melanins in the body. Their increased content is also observed in some malignant tumours-melanomas.

Obtaining melanin from fungi. Melanin from *Aspergillus rnger* was obtained according to the generally accepted method. Cultivation of the producer was carried out on mineral medium containing glucose at 26 ^{O}C until the appearance of strong pigmentation of mycelium. The mycelium was separated by filtration, washed with distilled water and frozen at -16 °C; subjected to alkaline hydrolysis (0.1 M NaOH) followed by precipitation with concentrated HC1. The precipitate was centrifuged at 7000 g for 30 min, washed with distilled water, centrifuged again and dialysed. Dialysis was carried out for at least 24 h and was terminated when the dialysed solution reached pH = 4.5. The dialysed fungal melanin preparation was dried $^{n}p_{I}$ 4° °C.

Using the same methodology, melanin pigments were obtained from the bazndnomntsets *Inonotus obUqum* (trout or chaga), *Phellmus robustas* (false oak trout) and *Phellmus ignianus.* The purified melanin pigments can be used to create pharmacological preparations with a variety of protective properties, e.g. the anti-inflammatory and anti-ulcer drug Befungin is obtained from birch chaga, a sterile form of the fungus ínonotusob/iquus.

Method of microbiological production of melanin. The invention relates to the field of microbiological synthesis of organic compounds. The method involves cultivation of a yeast strain *Saccharomyces neoformans* var. *nigricans* on carbon-containing media at 27...35 ^{O}C and pH of the culture liquid 4.5-5.5 under aeration and stirring. The obtained biomass is separated from the culture liquid with subsequent isolation of the target product. The method allows reducing the synthesis time and increasing the melanin yield.

Melanins are black, brown and yellow pigments. Melanin molecules are complexes formed by polymers of tyrosine derivatives and proteins. The molecular weight of melanin depends on the method of its production.

Here is an example of melanin production.

1. Yeast strain *Saccharomyces neoformans* var. *nigricans* GSP-TB 17D

were grown in an apparatus, the design of which provided aeration of the culture liquid with stirring and temperature control, with a volume of 5 dm^3 on nutrient medium, containing, as a carbon source, alcohol production waste in the amount of 4 % of the culture liquid volume, and as mineral nutrition sources ammonium sulphate (6 $g/dm^{3)}$, single-substituted potassium phosphate (1.5 g/dm), magnesium sulphate (0.5 g/dm), manganese sulphate, sulphate o
zinc, iron (II) sulphate and copper sulphate (0.05 g/dm°); pH of the culture medium was 5 5 at 30 °C, aeration mode - 1 ml gas/ml of culture medium - min. A suspension containing 3.9 % of biomass was obtained during 20 h of cultivation. The biomass was treated with 0.5 M sodium hydroxide at 0.5 atm pressure for 30 min. Yeast cells were precipitated by centrifugation. The f-gate was acidified with 40% hydrochloric acid to pH = 3.2. As a result of the above operations, melanin was precipitated. The precipitate was centrifuged at 5,000 rpm for 15 min, followed by successive washing with water and ethyl alcohol. 4.5 g of melanin was obtained, yielding 2.5 % of the substrate and 3.1 % of the biomass content.

2. Yeast *of Saccharomyces neoformans* var. *mghcans* was grown in a 5 dm^3 apparatus, the design of which ensured aeration of the culture liquid with stirring and thermostatting, on a nutrient medium containing as a carbon source a mixture of molasses and hydrocarbons in the ratio of 10. 1 in the amount of 6 per cent of the culture liquid volume:1 in the amount of 6 per cent of the volume of natural liquid, as well as mineral nutrition sources ammonium sulphate (5.8g/dm), single-substituted potassium phosphate (1.6g/dm), potassium sulphate (0.6g/dm), manganese sulphate, zinc sulphate, iron (II) sulphate and zinc sulphate PO0.045 g/dm, pH of the culture medium was 5.6 at 30 °C, aeration regime -1.1 ml gas/ml of culture medium - min. A suspension containing 3.5 % biomass was obtained during 18 h of cultivation. The biomass was treated with 0.5M potassium hydroxide at 0.5 atm for 30 min. The yeast cells were precipitated by zeitrifugation at 5,000 rpm for 20 min. The fugate contained 70 % melanin 15 % proteins and 6 % lipids. As a result, the product yield was 13.8 % of the initial biomass, with pure melanin obtained as 6.9 % of the biomass content.

3. Yeast *of Saccharomyces neoformans* var. *mgncans* was grown in a 5 dm^3 apparatus, the design of which ensured aeration of the culture liquid with stirring and thermostatting, on a nutrient medium containing alcohol production waste as a carbon source in the amount of 4% of the culture liquid volume, and as mineral nutrition sources ammonium sulphate (6 $g/dm^{3)}$, single-substituted potassium phosphate (1.5 $g/dm^{(3}$)), magnesium sulphate

(0.5 g/dm$^{3)}$, manganese sulphate, zinc sulphate, iron (II) sulphate and copper sulphate 0.05 g/dm each; The pH of the culture medium was 5.5 at 30 °C, and the aeration regime was 1 ml gas/ml culture medium ■ min. A suspension containing 3.3 % biomass was obtained during 18 h of cultivation. The biomass was treated with ultrasonic vibrations of 50-100 kHz for 5 min. Yeast cells were precipitated by centrifugation. The fugate was passed through a *Milhpore* membrane filter with a pore size of 0.45 µm. The obtained product contained 98 % melanin.

4. Milk whey was used as a carbon-containing raw material for melanin production by microbiological synthesis. Yeast strain *Saccharomyces neoformans* var. *nigricans* was grown in an apparatus, the design of which provided aeration of the culture liquid during stirring and thermostatting, with the volume of Yudm. on milk whey with the dry matter content of b %. Ammonium sulphate (4 g/dm), single-substituted potassium phosphate (1.0 g/dm$^{3)}$, copper sulphate (0.1 g/dm$^{3)}$, iron (II) sulphate (0.01 g/dm) were used as mineral sources; pH of the culture medium was 5.5 at 30 °C, aeration mode - 1 ml gas/ml of culture medium - min. A suspension containing 2.6 % biomass was obtained during 20 h of cultivation. To extract melanin, the biomass was treated in an ultrasonic disintegrator at an oscillation frequency of 100 kHz for 30. min, then the disintegrated yeast cells were precipitated by centrifugation. The fugate was acidified with hydrochloric acid to pH = 3.0 and the precipitate was separated by centrifugation. The obtained product contained 91% melanin.

As a result of the experimental work carried out, it was found that the molecular weight of the obtained melanin depends on the experimental conditions.

Chapter 17

BASIC PRINCIPLES OF ALKALOID PRODUCTION

17.1. Definition of alkaloids

Alkaloids are heterocyclic nitrogen-containing substances of alkaline nature with strong physiological action. In the second half of the XVIII century and in the beginning of the XIX century, when studying the chemical composition of plants, complex derivatives of heterocycles were singled out separately, which later received the unifying name "alkaloids" (Latin *alkali* - *alkali*, *oid.es* - similar, i.e. similar to *alkali*). The term was introduced by W. Meissner in 1819. The modern definition of alkaloids was first given in 1910 by E. Winterstein and G. Trier. The definition formulates the basic conditions that a true alkaloid must fulfil:

. plant origin;
. the complex molecular structure of the compound;
. atomazote-hour heterocyclic system;
. the manifestation of high pharmacological activity.

At present, the classical Winterstein-Trier definition is somewhat outdated, as the compounds considered as alkaloids by most chemists and pharmacologists do not fulfil all its requirements. This is not surprising, since more than 5 thousand alkaloids are known, and the structure has been established for about 3 thousand.

Let us draw attention to some disadvantages of the Winterstein-Trier definition: many substances with the structure of classical alkaloids are obtained from materials^ of non-plant origin (bacteria, fungi, animal tissues). The definition "complex molecular structure of the compound" is too vague and does not give an idea of the position as a whole. The phrase "high pharmacological activity" is not too correct, as the pharmacological activity of substances largely depends on the dose level and is manifested at sufficient content by many substances unrelated to alkaloids. Taking into account all the above-mentioned, a new definition of the concept "alkaloid" was proposed, which certainly covers a larger number of compounds, nevertheless excluding such classes of natural nitrogen-containing compounds as amino acids, amino sugars, aliphatic amines, proteins and peptides, nucleotides, nucleic acid, vitamins, pterins, porphyrins. The new definition formulated by W. Pletier, was that an alkaloid is a cyclic organic compound containing nitrogen in the negative oxidation degree and having an organic distribution among living organisms. This definition met the necessary conditions and gained widespread acceptance. The merit is that it confirms the classification as alkaloids of most of those compounds that are traditionally considered alkaloids but do not fit the classical Winterstein-Trier definition: for example, colchicine, piperine, ß-phenylethylamines, ricinine, genzianine, bufotoxin.

17.2. Methods of alkaloid extraction

The main problem of alkaloid extraction is the separation of the substance from the "ballast" material that constitutes the bulk of the plant material. In the case of volatile alkaloids, separation is done by distillation with water vapour, but this is rare. As the alkaloids are usually found in the plant in the form of various acids, it is necessary first to liberate them by wetting the crushed plant with an alkali solution.

Extraction with suitable solvents is often used for the extraction of alkaloids. Such techniques fall into two groups: extraction in the form of salts and extraction in the form of free bases.

Extraction of alkaloids in the form of salts. The plant raw material is treated with a suitable solvent to which a small amount of an acid (acetic, hydrochloric, tartaric, citric, etc.) is added. The extraction is usually carried out in conical extraction apparatuses (percolators), into which finely threshed plants are loaded.

The solvent is poured into the percolator. After infusion for several hours, the solution is slowly drained through a tap at the bottom of the percolator, and the raw material is again poured into the percolator with fresh solvent until it is completely extracted, i.e. until the presence of the alkaloid can no longer be detected by suitable qualitative reactions in a sample of the liquid flowing from the percolator. Continuous percolation is more perfect; in this method, as the solution is slowly drained from the tap of the percolator, the same amount of fresh solvent is automatically added to the upper part of the percolator. If the equipment is available, it is advisable to carry out extraction in several percolators on the principle of countercurrent: the solution flowing from the first percolator, goes to the fresh raw material located in the second percolator; from the second enriched solution goes to the fresh raw material in the third percolator, etc. As a result, it is possible to obtain more concentrated solutions of the alkaloid and to dispense with less solvent. In production, "extraction batteries" consisting of 5-10 percolators are installed.

Salts of alkaloids are usually soluble in water, alcohols (methyl and ethyl) and insoluble in ether and hydrocarbons, therefore, when extracting alkaloids in vidsolei "as a solvent is usually used water or alcohol. Although the extraction of alkaloids is easy and fast in most cases, however, this method has a disadvantage: in alcohol, and especially in water, a large number of so-called extractive substances (proteins, resins, tannins, mucilages, etc.) are extracted from plants along with alkaloids, the presence of which often makes the processing of such solutions difficult.

Extraction of alkaloids in the form of free salts. It is necessary to isolate the alkaloids in the plant in the form of salts beforehand, which is achieved by treatment with alkali. Sometimes, for this purpose, a slightly damp powder of plant material is thoroughly mixed with a dry base (magnesium oxide or lime) and then subjected to extraction. In other cases, the plant is wetted and thoroughly rubbed with an alkali solution (ammonia, soda, sodium hydroxide) and then subjected to extraction in a percolator. Since free alkaloids are soluble not only in water and alcohol, but also in most organic solvents, the choice of a suitable solvent in this case is much richer. Most often benzene and dichloroethane are used; less frequently ether, chloroform, hydrocarbon tetrachloride, petroleum ether, and paraffin are used. The extraction itself is carried out by percolation in exactly the same way as in the case of extraction in acid medium.

The choice of a suitable alkali is an important point, as many alkaloids are very sensitive to the action of strong alkalis and may undergo undesirable changes. Sometimes the alkaloid is such a strong base that weak bases like ammonia are not sufficient for its isolation from salts.

Pre-extraction. To separate a mixture of alkaloids from ballast substances, a method of preliminary purification of raw materials is used. For this purpose, the raw material is first treated with a weak acid (or a salt having a weakly acidic reaction) and subjected to extraction with benzene or petroleum ether. Alkaloids bound in the form of salts do not pass into these solvents, and the solvent extracts only neutral or acidic extractives. After preliminary purification, the plant material is again treated with a suitable alkali and extracted again by the same method. The solution of alkaloids is thus obtained more pure, containing much less extraneous substances, and the extraction of pure bases from it is much easier. However, owing to the cumbersome and time-consuming nature of this method, it is used only in exceptional cases where it is necessary to deal with raw materials particularly rich in ballast substances or in the case of very sensitive and easily changeable alkaloids.

The extracts obtained in one way or another contain alkaloids (and ballast substances) either in the form of salts or free. Accordingly, their further processing is somewhat different.

The extracts are further processed. In this case, for the isolation of alkaloids from aqueous, acidic solutions of their salts, these solutions are alkalised and the alkaloids are sucked off (if they are difficultly soluble in water and directly precipitate in solid form) or extracted with a suitable solvent (ether, chloroform, benzene, amyl alcohol, etc.), which does not mix with water. This treatment is often carried out repeatedly.

In the case of alcoholic acid solutions it is necessary to remove the alcohol first, which is done by distillation in a water bath; the thick mass remaining is then treated with water (or dilute acid), with some of the resinous matter remaining undissolved and separated by filtration. These resins often adsorb a considerable amount of alkaloids, so that it is necessary to treat them several times with hot water (or dilute acid) until the alkaloids are completely extracted from them.

Recently, a simpler adsorption method has been used for the isolation of alkaloids from aqueous or acidic diffusion juices.

Charcoal and ion exchange adsorbents: natural clays or artificial resins are commonly used as adsorbent.

Solutions of free alkaloids in a solvent not miscible with water, obtained by alkaline extraction of the plant, are usually much purer than aqueous and alcoholic extractions, i.e. they contain fewer ballast substances. To obtain alkaloids from them, these solutions are first shaken with dilute acid (1-5 %), into which all alkaloids are transferred. The latter are thus immediately concentrated in a relatively small volume of liquid. This acidic solution is subjected to the usual purification, alkalised, and the alkaloid mixture is either sucked off or extracted again with an organic solvent

Separation of alkaloids on the basis of different boiling points. This method is used when the alkaloids in a mixture differ greatly in their boiling points. It is thus possible to separate them by crushing distillation

Methods based on solubility differences. The difference in solubility of alkaloids and their salts in different solvents is the basis for the most frequently used methods of their separation and purification. Already in the extraction of the "sum of alkaloids" from the primary acidic solution obtained by extraction, a partial separation of the mixture can be achieved by using different immiscible organic solvents. Thus, for example, when shaking the alkalised solution with ether, "the latter transfers part of the alkaloids, while part remains in the aqueous solution and is extracted from it only when another solvent, such as chloroform or benzene, is used. This partial separation of an alkaloid mixture into two or more groups has been used, for example, in the case of the alkaloids of *Anhalonium* cactus. This separation is, of course, never complete and represents only a first rough fractionation.

Separation. on the basis of different "strength of basicity". The method is

based on the fact that different alkaloids have different "strength of basicity". If to a mixture of such alkaloids an amount of acid insufficient to neutralise the whole mass is added, the strongest bases will bind first to the acid, while the weaker ones will remain free. On the contrary, if to a solution of a mixture of alkaloids in a theoretical quantity of any acid be added
the amount of alkali insufficient^ to liberate the whole sum of alkaloids, the salts of the weakest bases will be decomposed first, while the stronger ones will remain in the acid-bound form. Separation by this method is usually incomplete, especially in complex mixtures, and "individual fractions observed enrichment of one of the bases. For complete separation and purification, this method is combined with other methods based on the crystallisation of co-leu or free bases.

Separation based on different adsorption capacities (chromatography). The chromatography method consists of passing a test solution containing several alkaloids through a column filled with adsorbent . After the solution has completely penetrated the adsorbent layer, the column is washed with an organic solvent or a mixture of several solvents and the individual fractions of the liquid flowing from the column are collected. Further routine processing of the individual fractions allows the isolation of individual compounds.

Separation by obtaining derivatives of alkaloids. This method is used when the alkaloids in the mixture differ in chemical features such that they can be separated by obtaining any suitable derivatives. This method of separation is based on the fact that one of the alkaloids reacts with a reagent while the other remains unchanged. The properties of the derivative of the first alkaloid thus formed (solubility, etc.) are often very different from those of the parent substance and allow its separation by the usual method of crystallisation. The main condition in this case is that the alkaloid should be easily obtained back from the derivative and should not undergo any significant changes.

Chapter 18

FERMENTATION PRODUCT RECOVERY

18.1 Lactic acid fermentation

In lactic acid fermentation, the end product is lactic acid, which people have known for a long time. Fermentation of milk, making sourdough, kefir, souring vegetables are the results of lactic acid fermentation of milk sugar or plant carbohydrates. This type of fermentation is carried out with the help of lactic acid bacteria, which are divided into two large groups (depending on the nature of fermentation): *homofermentative*, forming from sugar only lactic acid, and *heterofermentative*, forming in addition to lactic acid alcohol, acetic acid, carbon dioxide.

Homofermentative lactic fermentation is caused by bacteria of the genus *Lactobacillus* and Streptococcus. They can ferment various sugars with six (hexoses) or five (pentoses) carbon atoms and some acids. However, the range of products digested by them is limited.

Lactic acid bacteria have no enzymatic apparatus to utilise air oxygen. Oxygen is either indifferent to them or inhibits their development

Lactic acid fermentation can be described by the equation

$$C_6H_{12}O_6 \rightarrow 2CH_3 \cdot CHOH \cdot COOH + 21{,}8 \cdot 10^4 \text{ Дж.}$$

Heterofermentative lactic acid fermentation is a more complex process than homofermentative fermentation: the digestion of carbohydrates leads to the formation of a number of compounds that accumulate depending on the conditions of the fermentation process. Some bacteria produce ethyl alcohol and carbonic acid in addition to lactic acid, others acetic acid; some heterofermentative lactic acid bacteria can produce various alcohols, glycerol, mannitol.

Heterofermentative lactic acid fermentation is caused by bacteria of the genera *Lactobacterium* and *Streptococcus.* The chemistry of these fermentations is not as well understood as that of alcoholic or homofermentative lactic acid fermentation.

Hetero-fermentative bacteria form lactic acid in a different way. The final step, the reduction of pyruvic acid to lactic acid, is the same as in the case of homo-fermentative fermentation. But pyruvic acid itself is formed by a different cleavage of glucose - hexose monophosphate. The energy yield is much lower than in alcoholic fermentation

Hetero-fermentative bacteria digest a limited number of substances: some hexoses (of a particular structure), pentoses, sugar alcohols and acids.

Lactic acid fermentation is widely used in the production of dairy products:

sourdough, acidophilin, cottage cheese, sour cream. In the production of kefir, koumiss along with lactic acid fermentation caused by bacteria, there is alcoholic fermentation caused by yeast. Lactic acid fermentation occurs at the first stage of cheese making, then lactic acid bacteria are replaced by propionic acid bacteria.

Lactic acid bacteria are used in fruit and vegetable preservation and fodder silage. Pure lactic acid fermentation is used to produce lactic acid on an industrial scale.

Lactic acid is used in leather production, dyeing, laundry detergents, plastics, pharmaceuticals and many other industries, in the confectionery industry and for the preparation of soft drinks.

Typical, inotypical lactic acid fermentation. The causative agents of *typical lactic acid fermentation* are the following microorganisms: 1) *Bact. Streptococcus lactis* - small sporeless bacillus, young culture has the appearance of streptococcus; optimal temperature for the development of the bacterium is 30...35;C; it *ferments lactose, glucose, galactose and maltose; lactic acid is fermented* by *lactose*....35;C; it ferments lactose, glucose, galactose and maltose; lactic acid accumulates in the medium about 1 %; 2) *Vase Deìbriicki* is a long, sporeless bacillus that develops at a temperature of 45...48 °C; it is used in the production of lactic acid; 2) *Vase Deìbriicki* is a long, sporeless bacillus that develops at a temperature of 45...48 °C....48 °C; it is used for the production of lactic acid, the amount of which reaches 2.2 %; in the presence of neutralising agents (chalk) it is possible to accumulate up to 10 % of lactic acid; 3) *Baci, bulgancum*, or Bulgarian bacillus, is an immobile sporeless bacillus with a length from 10 to 20 μm, the optimum temperature for its vital activity is 40 ...48 ^{O}C. This bacterium accumulates the greatest amount of lactic acid. This bacterium accumulates the largest amount of lactic acid equal to 3.2 %. Lactic acid bacteria can digest different sugars depending on the presence of certain enzymes.

Atypical lactic acid fermentation is one in which other products such as ethyl alcohol, acetic and succinic acids, carbon dioxide and hydrogen are formed along with lactic acid.

The amount of products obtained varies. On average, lactic acid is formed up to 40 % of the fermented sugar, succinic acid - about 20 %, acetic acid and ethyl alcohol - U % each, and 20 % are gases. *B. coif, B. pentoaceticum* and others are representatives of bacteria causing atypical lactic acid fermentation. Pentoses are also subjected to atypical lactic acid fermentation. The fermentation of pentoses produces lactic and acetic acids:

$$C_5H_{10}O_5 = CH_3 \cdot CHOH \cdot COOH \cdot CH_3COOH.$$

пентоза молочная кислота уксусная кислота

pentose lactic acid acetic acid

There are also so-called false bacteria of lactic acid fermentation, they multiply at a temperature of 25...35 ^{O}C and form other acids along with lactic acid.

Typical and atypical lactic acid fermentation is widely used in various fields of national economy. The former is used for preparation of dairy products (sour milk, kefir, koumiss, etc.), sour black bread, production of lactic acid, preservation of fruits and vegetables. The fermentation of pentoses is of importance in the ensiling of fodder, as hay, straw and some plants contain pentosans hydrolysed to pentoses, which are converted into the above acids, giving silage a sour flavour and protecting it from rotting.

18.2 Preparation of propionic acid fermentation products (vitamin B_{12})

Vitamin Bj_2, or cyancobalamin, is an important biological compound, an active mammalian haematopoietic factor and a growth factor for many species of microorganisms and animals.

The history of vitamin B_{12} began with research into malignant anaemia (pericyclosio- anaemia), a disease that affects mainly the elderly but is also sometimes seen in children.

Until 1926, this disease was incurable and usually ended in death. It produces abnormally large, underdeveloped and unstable red blood cells, the total number of which was significantly reduced (one and a half to two times). The disease also affects other fast-growing tissues, such as the schizoid membrane of the stomach (as a result, the secretion of HC1 stops) and nervous tissues. The central nervous system is often demyelinated, with impaired coordination of movements (ataxia) and psychotic disorders Intramuscular administration of 3b micrograms of vitamin B_{12} is sufficient to induce remission in a patient with pernicious anaemia.

In 1926, J. Minot and W. Murphy discovered that it was possible to cope with pericyosioi anemia by consuming raw or lightly fried liver at the rate of 1/4 kg per day. It was only 22 years later that the first red cobalt-containing crystals of vitamin B_{12} were isolated from the liver of cattle. It was later discovered that liquid fermentation media of bacteria could be a richer source of the vitamin.

The structure of vitamin B_{12} was first determined in 1956 by D. Hodge-Keene by X-ray diffraction. It was the largest organic molecule whose structure could be determined by X-ray structural analysis. The cyclic system of vitamin B_{12} is similar to the cyclic system of porphyrins; it consists of four pyrrole rings with a cobalt atom in the centre

Normal blood levels of vitamin B_{12} are ~2. Milne is slightly higher, but vegetarians may have levels below half this value. Folic acid deficiency can also cause megaloblastic anaemia, and a large excess can reduce anaemia to some extent in patients with malignant anaemia.

In animal tissues, the concentration of vitamin is very low (1 mg/kg in bovine liver) to utilise this source for industrial purposes. Activated sewage sludge contains 4-10 μg/kg of vitamin, but this requires separation of a large number of different forms (analogues).

The chemical synthesis of vitamin B_{12} is very complex. It took more than 10 years for the staff of two large laboratories - D. Woodworth and O. Eschenmoser - to carry out the chemical synthesis of vitamin B_{12}, including 70etadia. Therefore, at present, vitamin $B_{(12)}$ is produced in industry exclusively by biosynthetic means. Of the 10 tonnes of vitamin B_{12} produced annually in the world, 3.5 tonnes is cyancobalamin (vitamin B(12) proper), 2 tonnes is hydroxocoba-lamine, 1 tonne is coenzyme B_{12} and a small amount is methylcoba-lamine; these forms are used in medicine in these quantities. The rest of the vitamin is used in animal husbandry.

Vitamin B_{12} is used in the treatment of pernicious anaemia, liver cirrhosis, nervous and mental disorders. It is widely used in feed production. Currently, most compound feeds for pigs and birds are enriched with vitamin B_{12}, especially favourable effect on animals is the combination of vitamin B_{12} with small doses of antibiotics, in particular biomycin. Vitamin B_{12} affects hematopoietic function and protein metabolism, takes part in the regulation of the optimal content in the animal body methionine, valine, threonine, leucine, isoleucine.

In poultry, reduction of B_{12} content in yolks leads to a sharp drop in hatchability of offspring. When B_{12} is added to feed (10-15 mg/t), the growth of piglets increases by 10-15%, chickens - up to $^{20\%}$, egg production of hens - from 180 to 208 eggs per year. When adding B_{12} to feed, it is possible to replace animal protein with vegetable protein. Domestic microbiological industry produces vitamin B_{12} of two grades: A - 200 mg/kg, B - at least 500mg/kg of preparation.

Humans get vitamin B_{12} from food and cannot utilise the vitamin produced by gut bacteria. Animals obtain vitamin B_{12} from food and utilise the vitamin produced by intestinal microflora.

Molecular structure. Vitamin B_{12} is the first organometallic compound isolated from a biological system. Of the nonpolymeric organic compounds,

it has the most complex structure. The molecule consists of two almost planar cyclic structures _|_O
and a linear region. The metal Co is bound by a macrococcal strongly resembling the porphyrin core of heme. The second ring structure is a nitrogenous base, 5,6-dimethylbene-zimidazole (5,6-DMB). It is connected to the first ring system by a heterogeneous side chain consisting of [1]-amino-2-propanol (isopropanol), esterified phosphate 3-mononononucleotide, connected to the 5,6-DMB base by a Na-gly-cosidic linkage
The structure of vitamin B_{12} (Fig. 13) is not only very complex, but also contains some unusual parts: 1) a corrin structure, which was not previously known in organic chemistry (until the discovery of vitamin B_{12} in 1948 independently by S. Rix and D. Smith$_{); 2}$) a Na-glycosidic bond. Rix and D. Smith); 2) Na-glycosidic linkage, which is very rare in nature and found only in a few compounds containing ribose-3-phosphate; 3) 5,6-DMB, which belongs to unique compounds and is found in nature only as part of cobalamins.

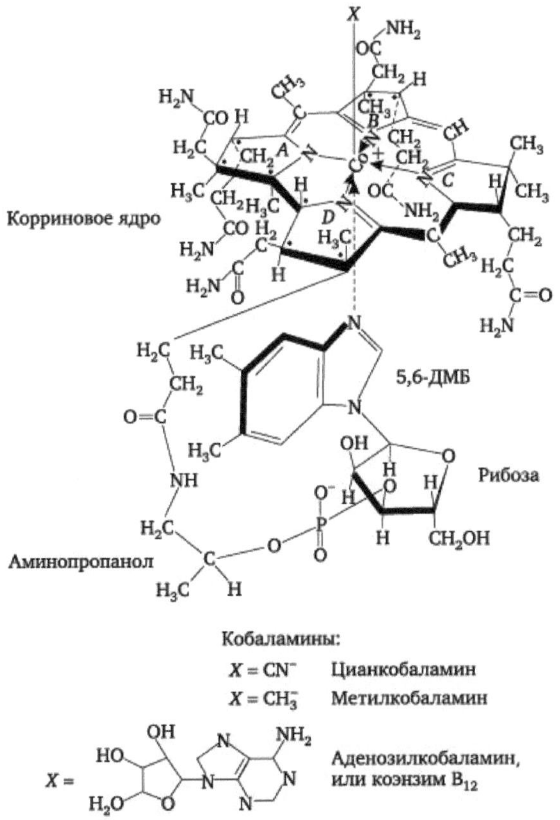

Корриновое ядро

5,6-ДМБ

Рибоза

Аминопропанол

Кобаламины:

$X = CN^-$ Цианкобаламин

$X = CH_3^-$ Метилкобаламин

Аденозилкобаламин, или коэнзим B_{12}

Cobalamins:

X = **CN** **Cyancobalamin**

X = $\mathbf{CH_3}$ **Methylcobalamin**

Adenosylcobalamin, or coenzyme $\mathbf{B_{12}}$

Figure 13: Structure of cobalamin

The cobalt atom has b coordination bonds: 4 of them are occupied by pyrrole rings,.one by H-3-5,6-DMB and the last one by the upper ligand (U), the nature of which may vary. In commercial vitamin B_{12} (cyancobalamin) the ligand is CN-rpynna (an artefact of the separation process).

The lower ligand.($_x$)-5,6-DMB can be replaced by.adenine (pseudo-dovitamin B_{12}), guanine (factor *C*), 2-methyladenine (factor *A*), and others. They may be active for some microorganisms but are inactive for humans. Of all the compounds similar to vitamin B_{12}, only Co-B_{12}-1 and Co-B_{12}-I (CH_3-$B_{(12)}$)) are active at the cellular level and, as cofactors, are involved in the catalysis of two types of reactionsAdenosyl B_{12} is used in reactions in which there is a rearrangement of carbon-carbon bonds. Methyl B_{12} is involved in methyl group transfer reactions, such as in the synthesis of methionine from homocysteine.

Biosynthesis. The biosynthesis of vitamin B_{12} involves steps of ob(formation :.

- of the porphyrin core;
- of the corrin nucleus;
- cobalamin

The pathway to uroporphyrin III is well studied in contrast to the subsequent steps from UPG P1 to cobiric acid.

In recent years, a number of pigmented intermediates, methylcorrinoids, have been discovered. The source of CH_3-groups is S-adenosyl-methionine. Dimethylcorrhythrin. is identical to syrohydrochlorin (the syro- heme of microbial sulfite reductase), hence this intermediate and possibly other corrhythrins have a vitamin-independent biolo- gical function

Cobalamy formation consists of amidation of seven carboxyl groups of the corrin ring, incorporation of an isopropylamino residue (derived from threonine), activation of cobipamide with the participation of GTP, and formation of a nucleotide base.

The incorporation of the cobalt atom into the system occurs shortly after the formation of cobiric acid. In the bacterium *Prop, shermami*, all known corrioids amidated more than cobiric acid are in the 5 -adenosyl form. Aerobic" aerotolerant forms synthesise 5,6-DMB from riboflavin via FMN. Oxygen is required for synthesis.

Anaerobes form 5,6-DMB in a different way, requiring glycine and methionine. 6-Aminolevulinic acid (6-ALA) is formed from glycine and succinyl-CoA in aerobes and from glutamate (as in plants) in anaerobes.

The biosynthesis of vitamin B_{12} in bacteria occurs by two independent pathways leading to the formation of 5'-deoxyadenosylcobinamide-guanosine diphosphate and a-ribazole-5'-phosphate, which are substrates for the final stage of synthesis.

Ribosomes capable of synthesising vntaminB_{12} in the reaction system were isolated from *E.* cob mutant cells. The main catalyser of synthesis is the 18L protein. 5S RNA proteins also catalysed the synthesis of the vitamin. Vitamin B_{12} synthesised with the participation of isolated ribosomes was bnologically active.

Haem has an inhibitory effect on the general pathways of corniond and porphyrin synthesis. The synthesis of vitamin B_{12} is regulated at the level of moyomethylcorphyrin. It is regulated by cobalamin and analogues in which 5,6-DMB is replaced by adenine or methyladenine. Factor B has no regulatory activity. The biosynthesis of vitamin B_{12} by propionic bacteria is stimulated by the NH^{4+} ion; in its absence, cells do not synthesise the vitamin. Glutamine and asparagine can replace NH* Stnmu- lator of B_{12} synthesis in the microorganism *Ps. demtnficans* is beta-in(trimethylglycine).

Producers. Vitamins are synthesised by many bacteria. Yeasts and celiac fungi do not form cornions. In humans and animals, vitamin B_{12} is synthesised exclusively by bacterial intestinal microflora.

Table 10

Formation of vitamin B_{12} by different strains

Strain	Carbon source	Yield, mg/l
Micromonospora sp. *Nocardia rugolaa*	Glucose	11,5
Propioníbacieriumfreudenreichii	Glucose cane molasses	14
Propioníbaderium Khermanii	Glucose	25
Propionibacterium vannielii	Glucose	23-40
Pseudomonas denitrificans Streptomyces oliváceos	Glucose	25
Bacterium FM-O2T	Sugar beet molasses	59
Methanobacillus omelianskii	Glucose	35
Protaminobacter ruber	Lactose	2,6
Corynebacterium	Methanol	8,8
Rhodopseudomon	Paraffins	2,5

Table 10 lists microorganisms that, due to the high level of cobalamin biosynthesis, have been considered at different times as its potential producers in industry. At present, mainly three bacterial strains are used here: *Pseudomonas demtnficans,* representatives of the genus *Propionibacterium* and methanogenic bacteria (mixed culture). In all cases, regardless of the strain used and cultivation conditions, cobalt ions and often 5,6-DMB are added to the medium. The addition of corrioid precursors such as glycine, threonine, 6-ALK and aminopropanol can have a stimulating effect on vitamin formation.

Production of vitamin B_{12} by propionic acid bacteria. At nowadays the following microorganisms are used for obtaining vitamin $B_{,2}$: *Prop,freudenreichu* ATCC 6207, *Prop,shermanu* ATCC 13673, *Prop,shermanü* BKM-103, their variants. The strains capable of independent synthesis of 5,6-DMB are of greatest interest. Since the synthesis of 5,6-DMB occurs better with access to air, a two-stage process is carried out, in which the highest yield of the product is obtained. At the first stage, the culture is grown under anaerobic conditions until complete utilisation of sugar. The second stage includes aeration, thus creating conditions for the synthesis of 5,6-DMB and the conversion of ethiocobalamin into deoxycobalamin. Both stages are carried out either in two different fermenters or in one. The anaerobically grown cells can be collected by centrifugation and the thick suspension incubated in air and, if necessary, in the presence of 5,6-DMB and cyanide. The addition of DMB is only made in the second stage of fermentation (if the bacteria do not synthesise it on their own), since the full forms of the vitamin, which inhibit its synthesis, are formed in its presence. Fermentation medium usually contains glucose or inverted molasses$_y$ (10-100 g/litre), large amounts of Fe, Mn and Mg salts and Co (10-100 mg/litre), sources of nitrogen. Corn extract (30-70 g/l) containing lactic and pantothenic acids, which enhance bacterial growth, is added to the medium. Pantothenic acid, which also stimulates the synthesis of vnta-min, is recommended to be added to the medium additionally. The bacteria are cultured at 30 °C, maintaining pH at 6.5-7.0 by adding $(NH_4)OH$. Fermentation is carried out in tanks containing 340 litres of medium inoculated with 7 litres of inoculum. In the first 80 h the culture grows under a low pressure of N_2 with weak agitation (without aeration), in the next 88 h aeration (2 m°/h) and agitation are included. Some variation in cultivation is possible. Vitamin B_{12} is retained in the bacterial cells, so its extraction is carried out as follows:

- release the vitamin from cells and convert it into cyanocobalamin;
- The crude product (80 % purity), which can be used in animal husbandry, is extracted;
- further purified to 91-98 % (for medical purposes). "DI™P^{ac}C™ ^{a}™ ^{ina} from cells are heated at 80...120 ^{O}C for 10-30 min at pH 6.1-8.5. Conversion to CN-ko-balamine is achieved by treating the hot solution or cell suspension with cyanide or thiocioate, often in the presence of $NaNO_2$ or B chloramine. Cormonoids are sorbed on various carriers: amberlite IRC-50, $A1_2O_3$, activated carbon - and eluted with aqueous alcohols or vodio-pheiol mixtures. From aqueous solutions, the corrn nonds are extracted with phenol or cresol or with a mixture of these alcohols with benzene, butanol, carbonic tetrachlorideomylichporoform.

The evaporation of various solvents yields precipitates or crystals of the vitamin, which are dissolved in a suitable solution of the required concentration.

The yield of vitamin B_{12} using propionic-acid by-products is 25-40 mg/litre. However, there is a patent report (France) on achieving an incredibly high yield of 216mg/litre.

Production of vitamin B_{12} using the bacterium *Pseudomonas*

denitrificans. A number of strains of the genus *Pseudomonas* produce significant amounts of *Bia*, but most often used mutant *Ps.* ***denitrificans***, which as a result of mutageeizaureaureaurea of vitamin B_{12} was raised from 0.6 mg/l (wild strain) to 60 mg/l. Bacteria were cultured with aeration and periodic stirring (or flow conditions) in the following medium: beet molasses - 100 g, yeast extract - 2 g, $MgSO_4$ - Zg, $MnSO_4$ - 200 mg, $CoNO_3$ - 188 mg, 5,6-DMB - 25 mg, $ZnSO_{(4)}$- 20 mg, Na_2MoO_3 - 5 mg, tap water - up to 1 litre, pH $7._4$. Melassa is rich in betaine and glutamic acid, which have a positive effect on vitamin output. Betaine stimulates the synthesis of b-ALA and possibly alters membrane permeability.

The culture is lyophilised and maintained in the above-mentioned medium. The culture is transferred into a test tube with dense medium. Medium composition: sugar-beet molasses - 60 g, brewer's yeast - 1 g $MgSO_4$ - 1 g, $MnSO_{(4)}$ - 200 mg, $ZnSO_{(4)}$- 20 mg, $MoSO_4$ - 5 mg, agar - 25 g, tap water - up to 1 litre, pH 7,4. Incubated for 4 days at 28 ^{O}C. Then cells are transferred into150 ml of liquid medium of the same composition (but without agar), placed in a litre Erlenmeyer flask. Incubate for 3 days at 28 ^{O}C on a rocker. The contents of the flask are transferred to a 5 L fermenter containing 3.3 L of medium (see above), sterilised for 75 min at 120 ^{O}C. Incubate for 90 h at 29 ^{O}C with stirring (420 rpm) o

and aeration (0.2 m°/h). Pure vitamin B_{12} is obtained through successive operations.

Vitamin B production by methanogenic bacteria. In the cells of *methanogenic* bacteria vitamin ^ is present from 4.1 nM/mg dry cells in *Methanosarcma barken* to 0.65 nM/mg dry cells in *Metanobacterium formicum.* The biosynthesis of cobalaminovarchebacterium (studied on M. *barken*) is similar to the biosynthesis of corrinoids in anaerobic eubacteria. In the *methanotroph Mtb. thermoautotrophicum*, most of the cellular cobamide is localised in the membrane fraction and bound to membrane protein. It is assumed that the integral membrane protein complex containing cobamide plays a ^{significant} role in the metabolism of these bacteria in the utilisation of H_2 $_+$ CO_2, which is apparently reduced to electron transfer. Corrinoids in methane-producing bacteria are also involved in the catabolism of methanol acetate. The conversion of methanol to methane in M. *barken,* occurs through the formation of CH_3-CoM, in the methylation of which two cobamide-dependent methyltransferases are involved for methanol. The corrinoid apparently serves as the prosthetic group of the enzyme.

In France, mesophilic metaiogeic bacteria were isolated from sewage sludge and incubated in association with other bacteria in a semi-flowing regime in a medium containing methanol (3-12 g/litre), molasses, corn extract, NH_4, Co, ortho-keylidia and 5,6-DMB. Fermentation is carried out at 35 ^{O}C in a 1,000 m^3 fermenter with daily replacement of 10% of the fermenting substrate with fresh medium. Biomass is separated on cena- rators and dried on a spray dryer. The dried concentrate is averaged with chalk to a standard activity of 1 thousand μg/g of the drug, which is used in this form as a feed additive. The dried concentrate before averaging contains 3 thousand μg/g of vitamin B_{12}, which is 45-50 % of the sum of corrinoids, factor III - 10-15 % and other incomplete corrinoids - 40-50 %. Thermophilic strains of *methanotrophic* bacteria of the genera *Metanobacillus* and *Metanobactenum* produce 2 mg/litre of cobalamin when the medium contains 8g/litre of methanol

1! In Russia, the production of vitamin B_{12} feed preparation is based on the processing of bard (waste from acetone-butyl or alcohol production) by a biocenosis of bacteria carrying out thermophilic methane fermentation of wastewater.

A complex consortium of anaerobic microorganisms including carbohydrate-, ammonifying-, sulphate-, and methane-forming bacteria is used. Methanol - up to 2 %, CoC_{12}-$6H_2O$ - 10 g/m^3, urea - 300 g/m^3 and dry fodder yeast - 230 g/m . Dosage of enrichments is performed automatically.

The bard is fed to the lower part of the digester fermenter (4-5 thousand $m^{3)}$, in which the process parameters are automatically adjusted, providing control of temperature, pH (7.5-8.0) and duration of fermentation. Fermentation is carried out continuously every day replacing 20-25 % of fermenting liquid with fresh bard. Fish oil is used as a defoamer.

To obtain a fodder preparation, the ferment is evaporated and dried. Since vitamin B_{12} is unstable during heat treatment, especially in alkaline environment, it is stabilised. For this purpose, the liquid obtained in the process of fermentation before evaporation acidify to pH 5.0-5.3 and add sodium sulfite (0.1-0.25%). The content of vitamin B_{12} in the initial fermented liquid is 4.4 g/m^3. Thickening of fermented bard is carried out on evaporators (up to 14-17 % of solids content), and drying - in a spray dryer. The concentration of vitamin B_{12} in the dried preparation is 500-600 mg/kg. True vitamin is 20-25 % of the sum of corrinoids, factor III - 35-40 %, factor B and others - 40-45 %. The resulting preparation is called BMP-12.

Mesophilic and thermophilic methanogenic bacteria, including *Metanobactenum therrnoautoOrophr Mb. therrnoforrnicurn, Mbbryanta, Metanosarcina barken, Ms. vacuolata., Ms. mazei, Methanococcus hatopilus*, synthesise exclusively factor III.

True vitamin B_{12} is formed by non-spore-forming *methylotrophs*:

Eubactenum limosum, the closely related *Butynbactenum methylotrophicum and Acetobacterwoodi.* By creating artificial biocenoses and selecting fermentation conditions, it is possible to regulate the process of vitamin B_{12} biosynthesis in a targeted manner.

New developments. In order to reduce the cost of vitamin B_{12} production and utilisation of cheap renewable raw materials, the formation of corrinoids by bacteria *Prop, ashbrgoryuta* ATCC 25562 was studied during growth on xylose as the main constituent of hemicellulose hydrolysates. Using xylose, the bacteria accumulated 0.35. mg of corrinoids in11 of medium without the addition of Co salt. For the production of corrinoids from xylose, a UFR reactor operating with ultrafiltration cell recirculation is most suitable.

Immobilised cells. In Japan, the stability and productivity of the biocatalyst were compared when the cells were incorporated into 240 y

Propionibacterium sp. into kappa-carrageenan, Na-alginate, arapa and fluoropolymer urethane gels. The optimal substrate was fluoropollnmer PU-9, whose polymeric matrix did not reduce the activity of the included cells. Under optimal fermentation conditions, 5 immobilised cells re-synthesised and excreted 900 μg of vitamin during 18 days of repeated periodic fermentation, demonstrating the possibility of multi-step complex synthesis (there are few examples of this kind known).

Improvement of producer strains. Improvement was achieved by mutation and selection. By this method the vitamin productivity of *Ps.derntnficans* was increased by 50 times. For Gram-positive bacteria *Propionibacterium, Bacillus, Streptotnyces,* protoplast fusion is applicable, for tramotritic bacteria such as *Pseudomonas*, conjugative plasmids such as Inc PI are available. So far, no significant practical results using these powerful methods have been obtained, but a beginning has been made. We cloned 11 genes encoding enzymes of vntamine biosynthesis in *V.* megatenum. *megatenum.* It is believed that the genome contains only 20-30 such genes, so the DNA of *Vas. megaterium* was subjected to fragmentation. *megaterium* was subjected to fragmentation and large fragments were incorporated into plasmids, which were then used to transform auxotroph mutants by B_{12}. Such mutants acquired the ability to synthesise vitamin B_{12}. The method can be used to produce strains in production scale.

Genes have been cloned in the bacterium *E.* soi. *Prop .techrncurn.*responsible for the synthesis of vitamin B_{12}. The bacterium *Prop, techmcum* does not contain plasmids, so DNA was isolated, purified and partially degraded from this strain, yielding fragments of 15-20 kilobases. These fragments were incorporated into the cleaved plasmid pVK322 and the resulting hybrid plasmid was transformed into *E. coli.* The new transformants differed from the control strain in morphological and physiological features.

18.3 Acetone-butyl fermentation

Acetone-butyl fermentation is a bnochemically more complex type of oil-acid fermentation, in which the fatty acids formed in the first phase are transformed into neutral end products - acetone, butyl *and* isopropyl alcohols. While in the first phase of fermentation the acidity increases, in the second phase the acidity not only does not increase, but even decreases. Thus, the pH of the medium (acidity) regulates the conversion of acids into neutral products. Acetobutylic fermentation of carbohydrates is caused by *Clostridium acetobutylicum* and closely related species. *CLbutylicurn* carries out fermentation very similar in type to acetone-butyl fermentation, but differs in that isopropyl alcohol is formed instead of acetone.

Acetone and butanol production. The production of acetone and butanol is one of the most important fermentation processes. It was first realised on an industrial scale by D. Weisman during the First World War. Acetone was necessary for the production of cordite and as an explosive in heavy artillery. Acetone of low quality was obtained by dry distillation of wood, but for the above-mentioned purposes a highhquality solvent was needed.

The fermentation process was based on the conversion of 3.8 % (weight/volume) starch by anaerobic spore-forming bacteria *Clostridium acetobutylicum.* Up to 30"/" of substrate was subjected to transformation, resulting in a solvent mixture (60% butanol, 30% acetone, 5-10% ethanol, isopropanol and mesityloxide). The rest of the substrate was converted into hydrogen and carbon dioxide during the fermentation process, as presented in the scheme (Figure 14).

Starch

Glucose
Glucose-6-phosphate
Triose phosphate
Pyruvate
Acetyl-CoA
Acetoacetyl-CoA
Acetone Butyrate
$_{D}$-2N^{+} T^{4N+}
Isopropanol n-Butanol

PUC.14. Schematic diagram of the reactions acetone-butyl fermentation

As large volumes of gases were formed, mixing was not necessary for large-scale production, and the main difficulty was to quench the foam. The acetone/alcohol ratio varied somewhat depending on the strain . Many starch-degrading and solvent-forming microbes can also digest molasses at sugar contents of up to 6 % (weight/volume). The sensitivity of the organisms involved in the process to n-butanol (upper limit of about 1.2 % by volume) and acetone (0.4 %) was found to be a factor determining the amount of substrate used.3 Infection of the fermentation tanks by microbial bacteria did not usually occur, and the main problem was infection by bacteriophages. Subsequently, it became clear that the micro-organisms involved in the process could be "immunised" by several resequences in the presence of bacteriophage. Phage infection was found to be strain-specific.

solvents were separated from the medium by distillation. At the end of the First World War, butanol production became a major player: it was used in the production of a wide range of substances, including urinary formaldehyde plastics, plasticisers and brake fluids. The by-product hydrogen was used in the production of synthetic methanol for the hydrogenation of edible oils; carbon dioxide was either liquefied or converted to dry ice. The waste solids contained high levels of riboflavnan (vitamin $B_{2)}$ and could be used as a protein-rich feed additive.

After the Second World War, fermentation production of these solvents was severely curtailed as the relative cost of petrochemicals compared to sugars and polymers decreased. The production of n-butanol by fermentation continued only in South Africa. However, nowadays butanol production by fermentation is becoming more and more favourable. The main disadvantage of the existing strains is low stability of the final products and relatively low solvent yields.

At present, acetone nbutyl alcohol is produced by digesting grain, molasses-grain mash or molasses. If the medium is prepared from grain, e.g. corn, then

firstly coarse flour is obtained, it is mixed with water at the rate of 6-8 kg of flour per UOl of water. Then mash boiled for 2 hours under pressure of 0.2 MPa and sterilised. Cooled to 37 ... 42 ^{O} C mass fermented for 2 days, pH medium 5-7. In the process of fermentationnzglucose is formed a mixture containingb parts of boo-type alcohol, 1 part of ethyl alcohol and 3 parts of acetone. Practically 1 kg of organic solvents is obtained from 3 kg of starch.
Methods of acetone and butyl alcohol production from sulphite alkali and wood hydrolysates were developed.
In acetone-butyl fermentation, acetic and butyric acids are formed in the first period, hydrogen and carbon dioxide are released. The butyric acid is then reduced to butyl alcohol. Acetone is formed from the condensation product of acetic acid, acetonacetic acid, by decarboxylation. During fermentation, a lot of riboflavin accumulates in the vereda, and its amount is greater the more intensely acetone is formed
Obtaining riboflavin. Methods of obtaining a feed preparation of riboflavin from acetone-butyl bard after separation of organic solvents have been developed. For this purpose, decanted acetone-butyl bard containing 2.0-2.5% of solids is used as a substrate. Fermentation takes place at 55...57 "C in non-sterile culture^ in two phases: in the first phase fatty acids and methane are formed, in the second phase - methane, carbon dioxide and vitamin B_2. The duration of the process in one apparatus is 2.5-3.5 days in two consecutive apparatuses - 2-2.5 days. The concentration of vitamin in the fermentation reaches 850 μg/litre. In parallel, gas (65 % methane and 30 % carbon dioxide$^{)}$is formed in considerable quantities, up to 20 m^3/m^3. The liquor has a slightly alkaline reaction. To stabilise the vitamin it is acidified with hydrochloric or phosphoric acid, then thickened to 20 % solids in an evaporator and dried in a spray dryer. The content of B_2 in the dry preparation is up to 100 μg/g.
Apparatuses for acetone-butyl fermentation. Since acetone-butyl fermentation is an anaerobic process, for fermentation purposes, apparatuses^ of anaerobic processes are used, which are quite simple and are used in the processes of conversion of plant raw materials, including plant wastes, as well as various industrial wastes. In methane fermentation for biogas production and in a number of other processes (acetone, champagne wine production) fermentation apparatuses (methane tanks) are used, which have different designs (from simple cesspools to complex metal structures or reinforced concrete structures) and volumes (from several to hundreds of cubic metres). Methane plants are equipped with a feed system, a pipe heat exchange system for temperature stabilisation, a simple mixing device for

homogeous distribution of feedstock and producer biomass, a gas hood and a variable volume device (gas holder) to collect the biogas produced.

Uses of acetone. Technical acetone is obtained by cumene method in the process of phenol synthesis, by acetone-butyl fermentation, and as a by-product in a number of industries. Acetone is used in the paint and varnish industry, in the manufacture of plastics, synthetic rubber and chemical fibres, and serves as a raw material for the synthesis of many other organic products. It is used in the production of acetic anhydride, acetone cyanohydride, diphenylolpropane, for degreasing surfaces, as a solvent in various industries. Prospects for the use of lower alcohols (methanol, ethanol), as well as acetone and other solvents as fuels for internal combustion engines have aroused great interest in recent years due to the possibility of their large-scale production in microbiological processes using various plant raw materials.

Industrial facilities for obtaining acetone-butyl fermentation products. Acetone-butyl fermentation from1t potatoes can produce 25 m° of hydrogen, 340 kg of butanol andIOkg of acetone.

Biohydrogen from biomass can be produced by butyl or acetone-butyl fermentation of sucrose or starch. In this process, 80 $m^{(3)\ of}$ hydrogen is produced from 1 tonne of molasses. This means that up to 140 m'm of hydrogen can be produced from sugar beet plantations (molasses), or in terms of all the molasses produced in 2003, 88 million metres of hydrogen. In addition to hydrogen, 1 tonne of molasses produces 114 kg of butanol and 36 kg of acetone, while the entire annual volume of molasses produces 125,000 tonnes of butanol and 40,000 tonnes of acetone.

Until the end of the 1970s there were 4 acetone-butyl plants in operation: in Grozny, Nalchik, Talitsa (Sverdlovsk Oblast) and Efremov (Tula Oblast). By the end of the 1990s, the Grozny and Efremov plants remained, and the Efremov plant could be restored. It produced up to 50 tonnes of solvents (butanol/acetone/ethanol - 13/4/1) and up to 29,000 m° of hydrogen per day, or 15,000 tonnes of solvents and up to 8.7 million $m^{(3)\ of}$ hydrogen per year, while the Grozny plant produced 74 tonnes of solvents and 43,000 m° of hydrogen per day, or up to 22,000 tonnes of solvents.

and 12.9 ml m of hydrogen per year. Unfortunately, all the hydrogen produced at that time was released into the atmosphere (the carbon dioxide was used ^{to} produce liquid and solid carbon dioxide).

In 1967 at Efremov ABZ and in 1969 at Grozny ABZ two shops for production of feed vitamin B_2 by thermophilic methane fermentation of liquid waste of these industries - bard (3 thousand m^3/day) were put into operation. In addition to vitamin B_2, each shop produced up to 30 thousand m^3 of biogas per day, which was entirely used to produce heat energy for the entire production cycle.

In the late 1960s, industrial production of biofuels from biomass (biohydrogen, biomethane, biobutanol, biobutanol, bioacetonebioethaiol) was established

In the Republic of Belarus "at present the producer of acetone and butanol is the plant JSC "Belchem", but there are sources of molasses and other raw materials that can be used in the microbial synthesis of acetone-butyl fermentation products.

18.4 Alcoholic fermentation

Ethyl **alcohol** (ethanol) is a colourless, easily evaporating liquid. Alcohol containing 4-5 per cent water is called rectified alcohol, and alcohol containing only a fraction of a per cent water is called absolute alcohol. Such alcohol is obtained by chemical treatment in the presence of water-removing agents (e.g. freshly calcined CaO).

Ethyl alcohol is a multi-tonnage product of the chemical industry. It is obtained by various methods. One of them is alcoholic fermentation of substances containing sugary substances in the presence of enzymes (e.g. zymase, a yeast enzyme). Such alcohol is called food or wine alcohol.

Ethyl alcohol can be obtained from pre-hydrolysed cellulose. The resulting glucose is then subjected to alcoholic fermentation. This alcohol is called hydrolysed alcohol.

As it is known, there are also synthetic methods for ethyl alcohol production - sulphuric acid or direct hydration of ethylene. The cost of alcohol obtained by this method is much lower than that obtained from food products.

Ethyl alcohol is widely used in various fields of industry, primarily in the chemical industry. It is used to produce synthetic rubber, acetic acid, dyes, essences, photographic film, gunpowder, and plastics. Alcohol is a good solvent and antiseptic, so it is used in medicine and perfumery. Ethyl alcohol is used to produce liqueur and spirits.

Ethyl alcohol used for technical purposes is deliberately contaminated with foul-smelling substances. Such alcohol is called denatured alcohol (the alcohol is coloured to distinguish it from pure alcohol).

Peculiarities of ethyl alcohol technology. Until the early 1930s ethyl alcohol was obtained exclusively by digestion of carbohydrate-containing raw materials and grain processing (rye, barley, corn, oats, millet). From 1930 to 1950, several methods of syn- tezaethyl alcohol production from chemical raw materials were discovered

In the USA in 1976, about 800,000 tonnes of ethanol were produced, including 550,000 tonnes by direct hydration (the rest by digestion of food raw materials). In the CIS, France and other countries ethanol is also produced by two-stage hydration of ethanol at 75...80 °C.

In.a.number.of.countries.ethyl.alcohol.is.obtained.by.digestion.of.hydrolysis. products.of.plant.materials. The purification of industrial alcohols is carried out by various methods. Raw food alcohol is usually freed from impurities (syrupy oils, etc.) by rectification. Syngue- tic ethyl alcohol is purified from ethyl ether, acetaldehyde and others by rectification in the presence of alkali and hydrogenation in vapour-phase nickel catalysts at 105 "C.

Recalcitrant alcohol is an atiotropic mixture of ethyl alcohol and water (95.57 % alcohol with a boiling point of 78.15 °C). Many applications require dehydrated, absolute ethyl alcohol.

Various sugars, such as grape sugar or glucose, have long been used to produce ethyl alcohol. Glucose is converted into ethanol through fermentation by enzymes (enzymes) produced by yeast fungi.

In its free form it is found, for example, in grape juice, fermentation of which produces grape wine with an alcohol content of 8 to 16 per cent

Polysaccharide - starch contained in potato tubers, rye, wheat and maize grains - can serve as a starting product for the preparation of starch. To transform into sugary substances (glucose) starch is preliminarily subjected to hydrolysis. For this purpose, flour or shredded potatoes are boiled with hot water and after cooling add malt-cross, and then dried and grinded with water barley grains. The malt contains diastase (a complex mixture of enzymes), which catalyses the saccharification of starch. At the end of saccharification, yeast is added to the resulting liquid, whose enzyme produces alcohol. It is distilled off and then purified by distillation.

At present, another polysaccharide, cellulose (fibre), which forms the main mass of wood, is also subjected to saccharification. For this cellulose is hydrolysed in the presence of acids (e.g. sawdust at 150...170 ^{O}C is treated with 0.1-5 % sulphuric acid under pressure of 0.7-1.5 MPa). The product thus obtained contains glucose and is fermented to alcohol by yeast. From 5500 tonnes of dry sawdust (waste from a sawmill of average annual productivity) it is possible to obtain 790 tonnes of alcohol (counting on 100 %-i). This saves about 3,000 tonnes of grain or 10,000 tonnes of potatoes.

Depending on the initial raw material and degree of purification alcohol-recti- fnkat divided into lux extra, higher purification and the first ™$r_{g;1}$.

Drinking alcohol 95 % is a mixture of rectified spirit of the highest purification and softened water, filtered and aged.

The rectified alcohol serves as a raw material for the production of distillery

products.
Vodka is a strong alcoholic drink containing 40-56% alcohol. It is made from rectified alcohol, softened water, filtered through activated carbon, added: granulated sugar, baking soda, vinegar, citric acid, honey, salt.
The distillery uses grain to produce ethyl alcohol from starchy raw materials. The main grain crops used in production are rye, wheat, barley and cu-"- - -.
The main indicator is the conditional starch content, which is used to calculate the yield of the starch.
When grain arrives, the grain admixture index is determined by the presence of other types of grain and it is determined whether it is a pure crop or a grain mixture.
The laboratory of the plant takes samples to determine the quality of grain in accordance with the requirements of GOST 13586-83 "Grain. Rules of acceptance and methods of sampling". When grain is transported by rail to check whether the quality of grain meets the normative requirements, the average sample from the incoming batch is analysed and a document on grain quality is issued. When grain is delivered by motor transport from collective farms, a note on quality is allowed in the delivery document.
From the moment the grain arrives at the plant's warehouses during the entire period of its storage, the quality and condition of each batch is monitored. The following indicators are used to assess the condition of grain during storage: temperature, humidity, and in batches of malted grain the energy and ability to germinate are additionally checked.
The incoming grain is weighed, cleaned and given to grain silage ($U = 48m^3$).
To produce rectified spirit, raw spirit is also used, which is delivered to the distillery by tanker trucks and stored in an open-type alcohol storage facility.
Rectified alcohol is produced at the plant by fermentation of starch- and sugar-containing raw materials with special yeast strains, distillation of the resulting alcoholic briquette and rectification at bragg-rectification plants, as well as by rectification of raw alcohol at a continuously operating rectification plant.
Ethyl alcohol rectified from food raw materials is produced in the form of alcohol-water mixture, the concentration of alcohol in which, depending on the grade, is 96.2-96.6 % by volume. In accordance with GOST 5962-67, rectified ethyl alcohol "Lux", "Extra", "High purification" is produced.
Super-Lux" alcohol is an innovation of RUE "Kristall" and is produced according to TURB0179 7526.542-97.
In addition to water, rectified ethyl alcohol contains various impurities (aldehydes, esters, higher alcohols and other chemical compounds) in very

small quantities, which form the flavour and aroma of the alcohol depending on the type of raw material processed.

According to organoleptic indicators ethyl alcohol and rectified alcohol must comply with the following indicators (GOST 5962-67):

- appearance - transparent liquid without foreign particles;

. the colour is a colourless liquid;

- taste and odour - characteristic of each type of ethyl alcohol produced from the respective raw material, without any aftertaste or odour of other substances.

The quality of water used for technological purposes has a great influence on technological processes and the quality of finished products. The most important indicators are hardness, acidity and bacterial purity. Water should not contain aquatic organisms visible to the naked eye, as well as extraneous harmful substances that inhibit yeast and deteriorate the quality of finished products. .

Depending on the place of use in the production process, different requirements apply to the water. Technical water with a temperature not exceeding 50 °C, a pH of at least 5.0 and a hardness not exceeding 12 mg/eq/l is used for mixing.

For the normal operation of the process, the water supplied for cooling to the heat exchangers and for vacuum cooling has a temperature $_{U}$ above 12 °C.

To feed the boilers, recycled water is used from the deflagrator with a temperature of 60...65 C. To prevent scale formation, the water is softened beforehand to remove calcium and magnesium ions. The same water is used for distillation of raw alcohol during rectification. Control is carried out by the factory laboratory.

Formalin (NSON) is a water-methanal solution of formaldehyde and is intended for use as an antiseptic.

Sulphuric acid (H_2SO_4) is used for acidification of yeast wort. The monohydrate content is at least 92.5 %.

Urea (H_2N-CO-NH_2, urea) is used in the plant as a nitrogen-containing food for yeast cultivation. The nitrogen content in terms of dry matter is 46.3 %.

Chlorine lime ($CaCl_{2)}$ is used for disinfection of malted grain, as well as for disinfection of fermentation vats, pipelines, various containers, floors and other contaminated places. Chlorine lime is a white powder with an active chlorine content of 32-35%.

The by-products of the digestion of fermentation are EAF, syrup $^{(oil)}$-$^{(oil}$)- .

Head fraction of alcohol TU RB 00966671-494-95 (change introduced on 10.09.1999) is a mixture consisting of methyl alcohol, acetic aldehyde, acids, esters of carboxylic acids and other impurities. It is used for obtaining technical and denatured alcohols "for various purposes.

The product of production, having passed the control shell, is sent for storage in the alcohol storehouse, in specially equipped metal tanks.

Syrup oil according to GOST 17071-91 - a mixture of amyl, nzobutyl, n-propyl and ethyl alcohols, water and insignificant amounts of other organic compounds - serves as a raw material for obtaining technical alcohols used in the food, perfume, pharmaceutical, paint and varnish industries.

During the distillation of the distillation process, intermediate products are formed in the distillery: distillate distillate, epurate, unpasteurised alcohol, syrupy alcohol, wash water from the ERC, bard and lauter water. With bard and lauter water the non-volatile part of the distillate is removed. Letu - 250 /

impurities are removed with the head fraction of ethyl alcohol, syrupy oils and distilled spirits.

Grain arriving from the warehouse is loaded into the receiving hopper equipped with a protective grid designed to separate large objects from the grain. On the spiral auger the grain is fed to the auger and enters the storage hopper of the dehulled grain. From there it goes to the air-sieve separator, where mechanical and organic impurities are separated by means of a sieve. The waste goes to a special hopper.

From the separator the grain is fed to the hopper of cleaned grain, and then to the weighing-doser to determine the amount of cleaned grain. From the weigher-doser the grain in portions of up to 50 kg is fed to the silo by elevators.

In the process of grain preparation dust is emitted, which can form an explosive mixture, therefore all augers, hoppers, burrows and air-sieve separator are equipped with aspiration, consisting of fans and cyclones, where the dust enters.

From the silos the grain is fed into the production storage bin.

Boiling of starchy raw material in continuously operating units is provided in the crushed form. The cleaned grain from the storage tank is gravity fed to hammer crushers and crushed there. The crushed grain is allowed to contain coarse residue (7-8%) with a 1 mm diameter.

The milled grain is continuously fed by gravity to the mixers. In the same place through rotameters is fed the required amount of water with a temperature of 40 45 C For the preparation of masha 1kg of milled grain takes 2.5-3.5 litres of water, depending on the starchiness of the raw material. The concentration of the wort at the correct dosage should be 16-18%

according to the saccharometer.

Grain mash by a plunger pump through a pipe-in-pipe heater, where it is heated by secondary steam from the holding tank to a temperature of 55...Ó5 °C, is fed to the contact head of the l-stage, where it is heated to a temperature of 65...75 °C. 65...75 °C. In the contact head of the 2nd stage the grain mixture is heated to the maximum possible temperature of 138.140 °C. When heating to a higher temperature, the amount of steam supplied to the 1st stage cooking column is reduced.

The cooking apparatus consists of one l-stage column and three 2nd stage columns. Uniformity of boiling of the mixture is achieved by the multi-sectionality of the apparatus and special design of the column of the l-i stage. Boiling time should be 40-50 min at temperature 138...140 ^{O}C. For maize it is recommended to cook at 144...150 ^{O}C for 60 min. In the cooking columns of the 2nd stage steam is not supplied. The cooking apparatus operates at the same pressure in the cooking columns l-i and 2-stage and the body of the level regulator, which is ensured by connecting them by an equalising line. The mass transfer is realised only due to the difference of levels in the columns.

The boiled mass from the last column of the 2nd stage is blown out of the vessel through the level regulator. In the level regulator the blowing out valve is located below the working level of the mass in the vessel by at least 500-600 mm. Non-condensable gases and vapours are periodically, after 15 min, during 15s, discharged into the steam pipeline through the circulation valve located in the upper part of the columns, and further into the vatmosphere.

A constant overpressure of about 0.02 MPa is maintained in the incubator, which corresponds to a kneading temperature of 105 ^{O}C. The mixture is kept in the incubator for 15-20 min. The uncooked mass is kept in the incubator for 15-20 min. The quality of digestion is determined visually, at least 4 times per shift. Sampling is carried out from a sampler mounted on the blowing tube.

The calculated amount of amylosubtilium is weighed or measured in litres and added to a consuming vessel containing water at a temperature of 20...30 ^{O}C. 10-15 minutes before the solution is fed into the saccharifier, the calculated amount of glucavamorin is added from the measuring tank under constant stirring.

Saccharification is the hydrolysis of digested starch to digestible sugars under the action of amylolytic enzymes, which contain saccharifying materials (enzyme preparations). At the same time, under the action of proteolytic enzymes, proteins are hydrolysed.

In continuous saccharification, a special installation is used to cool the pulp under vacuum. Before entering the saccharifier, the boiled pulp is passed through an evaporator in which the temperature of the pulp is reduced to 62 ... 63 °C because the vacuum in the evaporator is 0.08-0.081 MPa.

The vacuum in the evaporator is maintained by a sharp drop in temperature and condensation of the vapours released from the boiled mass. The vacuum in the evaporator is created by means of barometric condensers and a dry-air vacuum pump. The condensed vapours, together with the cold water supplied to the barometric condensers, are discharged into a collection tank, from where they are sent to the sewage system. The cooled mass from the evaporator enters the saccharifier, where it is mixed with enzyme preparations and quickly cooled to a temperature of 57...B8 °C. If the temperature does not reach the required temperature, additional cooling is carried out using the water jacket of the saccharifier. The duration of saccharification is not less than 10 minutesSalted milk is dosed by volume, its consumption is 16-18% of the volume of decoction. The enzyme preparation solution is dosed depending on the activity and starchiness of the saccharified mass. The sugared mass is centrifugal pumped through heat exchangers, where it is cooled to a temperature of 18...2O °C, in the brodnlnychan.

The main purpose of alcoholic fermentation is the digestion of the sugars of the wort by yeast into alcohol. The intermittent fermentation process is carried out from start to finish in a single apparatus. When filling sterilised and cooled fermentation vats, the necessary amount of yeast is introduced directly with the wort. Their consumption should be 6-8 % of the volume of fermented wort. The fermentation vat should be filled for no more than 8 hours. Longer filling times lead to a deterioration of the fermented wort. The duration of fermentation, counting from the beginning of filling the vat to the beginning of distillation of the mature fermentation, is 72 hours. The temperature during the three-stage periodic fermentation should be 18 ... 20 ^{O} C. During the main fermentation the temperature is 29...30 °C, during the pre-fermentation - 27...29 °C. Reducing the temperature during fermentation prevents yeast starvation and helps to prevent the growth of acidity of the fermentation. Temperature is regulated by supplying cold water to the water jacket fermentation vats and measured with a mercury thermometer in a protective sleeve, while the broth should not be overcooled, as it will slow down its fermentation. During fermentation fermentation vats are connected to the alcohol trap for condensation of spntartovyh vapours, carried away by the carbon dioxide released during fermentation. The wash water from the alcohol trap is taken to a collection tank, from where it is sent directly to the distillation unit. The concentration of alcohol in the wash water should not exceed 7 %o. The alcohol content in the wash water is determined with a glass alcoholometer. At the end of fermentation, the mature fermentation from the fermentation vats is sent to the distillation department. Fermentation vats are washed with antiseptic and steamed to a temperature of 105 ^{O}C after release from the fermentation. During distillation, the fermentation vats are completely emptied by means of plunger pumps.

The normative parameters of a mature fermentation are acidity, amount of unfermented carbohydrates and strength. The acidity increase in mature $^{1,|M}Λ^{№}$ T' under normal conditions of production should not exceed 0.15-0.20 deg. of acidity. The content of unfermented carbohydrates at excellent performance - up to 0,25g/100 ml; at good performance - from 0,25 to 0,35 g/100 ml; at satisfactory performance - up to 0,45 g/100 ml. Fermentation strength under normal production conditions should be 7.5-9.0 %o.

For operational control of the fermentation process in the processing of starchy raw materials determine the content of solids in the filter of the fermentation by saccharometer. Visible concentration of mature fermentation depending on the type of raw material should have the following values: maize - 0.4; millet - 0.2; rye - 0.7; barley - 0.7; oats - 0.9 %.

The aim of the process is to produce mature yeast grown on specially prepared yeast wort and used for fermentation in the fermentation vats.

The objective of the process is to conduct the process in such a way as to maximise the accumulation of yeast cells in the absence of the development of extraneous microorganisms.

For yeast production, wort is prepared from various cereal crops. Thick-skinned grains (barley, oats, millet) should be mixed with unmoulded grains so that cover fermentation is completely excluded.

In alcohol production, the process of yeast multiplication occurs at high acidity of the medium (pH 3.8-4.2), which inhibits the development of lactic acid bacteria and creates normal conditions for the development of yeast cells. Sulphuric acid is used to acidify the yeast wort.

When processing grain raw materials of normal quality, with the exception of maize, the yeast wort is prepared without additional malt nutrition. . When processing defective raw materials and maize, additional malt nutrition is added to the yeast wort. The technological instructions provide for the following norms of additional malt nutrition: 0.8 kg/dal of yeast wort - when processing maize; 0.6 kg/dal of yeast wort - when processing other raw materials.

The technological process of yeast cultivation consists of two operations: preparation of yeast wort and digestion of this wort in order to accumulate yeast cells in it, which are used as seed yeast for wort fermentation in fermentation vats.

Preparation of yeast wort. The yeast wort preparation tank must be thoroughly rinsed before the start of the process, disinfected with bleach solution for 254 *yr*.

25_ZO"IL"'3"aTeM B "PTM warm water and steamed at a temperature of 100 ^{O}C for at least 20 minutes. In this yeast cell the wort is taken from the clarifier after feeding in the yeast with all saccharifying material

The wort concentration should be between 17 and 18 %. This volume of wort is kept for 1 h for complete saccharification of starch, and then pasteurised at 80 °C, cooled to 50 °C and acidified with sulphuric acid from a measuring tank with careful stirring to 0.7-0.8 grad, acidity, which corresponds to pH 3.84.0. To acidify the yeast wort use sulphuric acid in diluted. form in compliance with all safety regulations. After cooling the yeast wort to a temperature of 30 ^{O} C in it cen- trobezheyymnasovym pump introduced mature yeast seeded yeast, in an amount of 8-12% of the wort volume. The wort is then cooled down to a crease temperature of 18 ... 2O °C.

Fermentation of yeast wort. The process of fermentation of yeast wort is carried out in yeast tanks with strict control over the temperature and purity of fermentation. Increase in fermentation temperature above 30., C is not DOPU(C)™, as it0 indicates the development of extraneous microflora.Yeast is considered mature upon achievement of visible otbroda, equal to 1/3 of the initial concentration of yeast wort.

Requirements for mature yeast: yeast cells should be well-nourished; the presence of dead cells should not exceed 3.0 %; foreign microorganisms should not be present in the microscope.

The technological scheme of the distillation unit with low vacuum under low vacuum of the BRUVAK-M type consists of four columns: distillation, epuration, rectification and extractive-rectification columns.

The liquor from the fermentation compartment is pumped to the rectification column heater, where it is heated to a temperature of 53...55 ^{O}C. Then the liquor goes to the distillation column heater, where it is heated to a temperature of 53...55 ^{O}C. Then the liquor goes to the distillation column heater, where it is heated to a temperature of 85...90 "Sparom, distillate. The heated digestate is sent to a separator, where it is freed from carbon dioxide. From there it is placed on the feed plate of the distillation column, and carbon dioxide in a mixture with other gases and water-alcohol vapours goes to the condenser of the distillation separator and further to the distillation column alcohol trap, where they are condensed, and the condensate is fed to the 36th feed plate of the evaporation column. Non-condensable gases from the alcohol trap of the distillery column are discharged to the atmosphere through a fire barrier and a breathing valve. In the distillation column, alcohol and other volatile components are extracted from the liquor. The bard is discharged from the column cube through a bard regulator into the bard collector.

Water-alcohol vapour from the upper bowl of the distillation column is transferred to the intertube space of the distillation column distillation heater. The distillate from it and its water part goes to the 28th feed plate of the epuration column. Part of the distillate from the aqueous part of the distillation column brew heater returns to the upper plate of the distillation column through the rotameter for its irrigation. The distillate distillate from the condenser of the distillation column and its alcohol trap goes to the 36th feed plate of the epuration column. The distillation distillate from the boiling pot of the evaporation column is collected in a collector, from which under the influence of the injector operation it is fed to the 23rd feed plate of the evaporation column. At the same time the level in the collector is kept constant.

Water-alcoholic vapour from the epuration column is fed to its deflegmator, condenser and alcohol trap for condensation. Phlegm from these heat exchangers is returned to the upper plate of the epuration column. The head fraction of ethyl alcohol is taken from the condenser of the evaporation column and through a refrigerator, a filter, a rotameter, an ejector and a spirit measuring device is taken to the alcohol receiving section.

The epurate from the cube of the epuration column is directed to the plate of the rectification column.

Alcohol vapours from the rectification column are sequentially condensed in the rectification column's liquor heater, its deflegmator, condenser and alcohol trap. Unpasteurised alcohol is taken from the condenser of the rectification column, the flow rate of which is controlled by a rotameter, and sent to the upper plate of the epuration column. The excess of unpasteurised alcohol is returned to the upper plate of the rectification column.

The rectified alcohol is taken from the 71-73, 75. 77 and 79 plates of the distillation column and through the alcohol cooler, epsuvette, alcohol measuring apparatuses is sent to the alcohol receiving department.

Luther water from the rectification column is diverted through a vapour separator through a vapour trap to the luther water collector, from which it is

pumped to the luther water pressure tank. From this collector luther water is fed through rotameter to the upper plate of the extractive rectification column and to wash the liquor oil in the decontractor nozzle column.
From the 5th, 7th, 9th, 11th zones of the lower plates of the cracking column the vapours of syrupy oil are taken, which are injected into the cube of the extraction and rectification column. The vapours are directed to the phlegmator and condenser of the extractive rectification column. Heterogeneous distillate from the deflegmator is directed to the separator, the lower layer of which is returned to the upper plate of the extractive-rectification column, and the upper layer, enriched with higher alcohols, is fed through the rotameter to the nozzle column of the decontractor, where it is mixed with luterne water. In the decontractor, the syrupy fraction rises to the top of the decontractor, from where the settled syrupy oil is collected in the syrupy oil collector. The cube residue from the extraction and rectification column cube and the syrupy layers from the decontractor are sent to the 12th plate of the distillation column or to the 7th and 9th plates of the rectification column.
From the 19th, 21st, 23rd and 24th zones of the distillation column plates, the syrupy alcohols are fed by means of an injector to the bottom plate of the evaporation column, and also periodically in the amount of 0.5 % of syrupy alcohols are taken to the alcohol-measuring shell of the EAF through a test refrigerator.
Vacuum in the epuration column and in the top of the rectification column is created by vacuum pump through barometric condenser, for which purpose the alcohol traps of epuration and rectification columns are connected by gas line with the bottom part of barometric condenser and further with vacuum pump. Cold water is fed to the barometric condenser to capture alcohol and other volatile substances, and the aqueous-alcohol liquid from it is directed to the 12th plate of the distillation column.
The vacuum at the top of the columns is regulated by supplying cooling water from the cold water collector on the heat exchange surfaces of the column heat exchange equipment. Hot water after the column heat-exchange equipment is discharged to the hot water collector. Steam condensate from the rectification column boilers is returned to the boiler house.
To control alcohol losses in bard and lauter water, the unit is equipped with sample coolers. To protect the columns from deformation due to overpressure or deep vacuum, the top and bottom of the distillation column, the bottom of the rectification column and the bottom of the extraction and rectification column are equipped with vacuum breakers.

During the processing of raw spirit, it is fed from the alcohol storehouse into the raw spirit collector, from which it is pumped to the pressure pump.

raw spirit collector. From there the raw alcohol through rotameters and mixer, where it is mixed with softened water of the pressure tank, goes to the raw alcohol heater, where it is heated to a temperature of 73...75 ^{O}C. The heated mixture of raw alcohol and softened water is fed to the 23rd and 28th plates of the suction column. The heated mixture of raw alcohol and softened water is fed to the 23rd and 28th plates of the separation column.

The rectified alcohol obtained at the distillation unit is discharged into storage tanks after being metered by alcohol-measuring apparatuses.

The design of the alcohol-receiving department provides for the possibility of discharging rectified alcohol into a separate tank for individual accounting and the possibility of controlling the discharge at any time. Samples are taken from the storage tanks in accordance with the established procedure for chemical and quality control of rectified alcohol. All storage tanks are connected at the top with overflow devices to prevent overfilling of the tanks with rectified alcohol and spillage losses.

To transfer the rectified alcohol from the storage tanks by gravity flow to the measuring tanks, where its volume and temperature are measured, and then to the alcohol storehouse for storage.

Chapter 19

ISOLATION AND PURIFICATION OF HORMONAL PREPARATIONS

Hormones (Greek: hormao - to set in motion, to excite) are biologically active substances of different chemical nature, formed by specialised cells of the glands of vitreous secretion, which are secreted directly into the blood and lymph and regulate metabolism and physiological functions of the organism. About 60 biologically active secretions are known, which are produced by endocrine glands and have hormonal activity.

Hormones are distinguished by their chemical structure:

- of protein nature: simple (insulin, prolactin, growth hormone) and complex (folatropin, lutropin, thyrotropin) proteins;
- peptide nature: glucogen, calcitonin, somatostatin, va- zopressi oxytocia;
- amino-alcohol derivatives: adrenaline, noradrenaline, oxytocin
- lipoid nature (steroid hormones): corticosteroids, androgens, estrogens, prostaglandins;
- tissue (parahormones): gastrin, secretin, heparin.

Endocrine hormones are obtained by chemical synthesis, genetic engineering and isolation from animal raw materials. There are also biotransformation techniques. For example, one strain of the bread mould *Rhizopus arrhizus* is used in the initial synthesis of the steroid derivative cortisone by hydroxylation of progesterone at the 11th position.

Insulin and some protein hormones of the pituitary gland (corticotropin, somatotropin), etc. have been synthesised. Preparations of steroid hormones, amino acid derivatives and peptides (oxytocin, vasopressin) are obtained by chemical synthesis. The production of polypeptide hormones is unprofitable due to multiple stages and complexity.

The following will give examples of how some hormones are obtained from various sources.

19.1. Insulin production.

Insulin isolation from animal raw materials

Insulin is a pancreatic hormone produced by the ß-cells of the islets of Langerhans and plays an important role in carbohydrate metabolism.^ The human insulin molecule consists of two poly-peptide chains, *A* and *B,* connected by two disulfide bonds. *The A-chain* comprises 21 and the *B-chain* comprises 30 amino acid residues.

In 1952, insulin was first isolated in crystalline form. Currently, there are several methods for isolating insulin from the pancreas of cattle and pigs.

The production of insulin consists of a number of stages:

- Crushing of frozen pancreatic glands and extraction with an acidic alcohol solution;

. deposition of ballast proteins and release of lipids;

- isoelectric precipitation of the insulin fraction (at pH = 5.5) and precipitation with alcohol, acetone, ether;
- Insulin purification: salt precipitation, fractionation by chromatography, gel filtration;
- precipitation of insulin in the form of crystals;

. zinc-insulin re-precipitation.

Fresh or frozen pancreatic glands are crushed (B) with a worm grinder and extracted by bnsmaceration first time with 80-85 % ethanol in a reactor with stirrer. The second extraction is carried out with 57 % ethanol acidified with orthophosphoric acid (hydrochloric or sulphuric) to pH = 2.8-3. Extraction is carried out for 1.5-4 h with constant stirring. Acidified alcohol contributes to the inactivation of the enzyme trypsin, located in the pancreas, due to which it is possible to keep insulin in an unchanged state. The endocrine drugs plant in Minsk uses a rotor-pulsation extraction apparatus, which makes extraction more intensive and reduces the time to 1.5 hours.

The resulting extracts are combined, left to stand for 48 h to free any undesirable proteins that precipitate.

The precipitate is separated by centrifugation and removed. Then ion-exchange chromatography is used for isolation and purification of insulin. Insulin is sorbed from a clear liquid on macroporous sulfocathnoniteKU-33-30/100 at pH = 3.0-3.3 in a fluidisation mode. Fat is removed by washing the cathnonite with 65-67 % ethanol, and ballast proteins - with 0.3 M acetate buffer solution. Insulin is desorbed using 0.01-0.05 M ammonium buffer solution (pH 10) and immediately acidified with hydrochloric acid to pH = 4.5 and acetone is added. The precipitated precipitate of ballast substances is removed. Insulin is precipitated with zinc acetate solution (pH = 6.2). Obtained zinc-nsulinopurify method crn- stallisation, then dissolved in water acidified to pH = 2.8 with citric acid. The solution is sedimented for 1 h. The precipitated ballast substances are removed by filtration through diatomaceous earth. The filtrate is mixed with acetone, tsnik chloride and phenol cooled to 0 C are added. For slow crystallisation of insulin, conditions are created with successive changes in the pH of the solution. The solution is alkalised to pH = 8.5, left for 2-3 min, change pH to 6.8, stir for 1 h; at pH = 6.5, stir for 2 h; pr" pH, equal to 6.2 and 6.0, - 2 h and stand for 20 h; at pH 5.8 - 2 h and stand for 96 h at 5 ^{O}C. Precipitated crystals of insulin are

separated by centrifugation, washed on a Buechner funnel with cold water, acetone, ether. Drying is carried out in air in a fume cupboard and desiccator.
The Danish company *Novo Industries* produces human insulin using a method based on the replacement of an alanine residue in the *B* chain with a threonine residue. This is achieved by enzymatic substitution followed by chromatographic purification of the product, resulting in a single-component human insulin with 99% pure drug content.
Creation of industrial production of genetically engineered insulin. The first attempt to create such production in the USSR can be considered a pilot plant of VNIIA, based on the technology developed jointly with the German firm *Genbiotech GmbH* in 19871989. The process was based on the production of recombinant protein (molecular mass of about 17 kDa) by a genetically engineered strain of E soy containing a leader amino acid sequence linked through an amino acid residue of methionine to human proinsulin. The initial technological scheme included the following stages:
- fermentation (growing the producer strain in a fermenter);

-Departments of biomass;
- cellular disruption with the release of inclusion bodies;
- * Recombinant protein was sequenced by ion-exchange chromatography on DEAE-Sepharose FF and gel filtration on SephadexC-25M;

- of recombinant protein cleavage by bromocyanine;
. oxidative sulphitolysis ofproinsulin;
- chromatographic purification of proinsulin hexasulfonate by gel filtration on Sephadex G-50 F, ion exchange and chromatography on DEAE-Sepharose FF and gel filtration on SephadexC-25;

. renaturationproinsulin;
- Chromatographic purification of native proinsulin by ion-exchange chromatography on DEAE-Sepharose FF;
- enzymatic cleavage of proinsulin;
- Chromatographic purification of insulin by ion-exchange chromatography on S-Sepharose FF, ultrafiltration and gel filtration on NephadexC-50 5C;
- freeze-drying of Na-salt of human insulin.

Despite the obvious cumbersomeness of the above scheme of the technological process, the VNIIA pilot plant produced pilot batches of genetically engineered human insulin (GICh) and the corresponding temporary pharmacopoeial article (TPA). However, with the collapse of the Soviet Union and the accompanying economic and organisational consequences, the production of the drug was discontinued.

In 1996, in accordance with the Decree of the President of the Russian Federation "On Measures of State Support for Persons with Diabetes Mellitus", the Federal Programme "Diabetes Mellitus" was adopted by the Government of the Russian Federation. On this basis, under the auspices of RAO "Biopreparat", large-scale research on the development of industrial technology and production of preparations of genetically engineered human insulin was resumed at the Institute of Bio-Organic Chemistry (IBC) of the Russian Academy of Sciences and the State Scientific Centre for Applied Microbiology (Obolensk). As part of the joint work, highly effective strains producing human preproinsulin were created and laboratory technology for production of insulin substances was developed. To date, CJSC National Bnotechnologies (the legal successor of RAO Bio-Preparation) in Obolensk has opened pilot production of human genetically engineered insulin substances, but there is no possibility to produce ready-to-use drugs.

The technological process of HCH substance production in IBCH RAS is carried out according to the following scheme:

- obtaining inoculum of the producer strain:
- revitalisation of the tinned culture;
- Growing a mother crop;
- inoculum cultivation;

- hybrid protein biosynthesis:
- cultivation of the producer in a fermenter (fermentation);
- biomass extraction (separation);
- disintegrating the lung suspension;
- isolation and washing of inclusion bodies (centrifugation);

. isolation and purification of recombinant protein:
- solubilisation of inclusion bodies and recovery of recombi- nant protein;
- Renaturation of recombinant protein;
- chromatographic purification of recombinant protein;

- Enzymatic cleavage of recombinant protein:
- Chromatographic purification of insulin 1;
- Chromatographic purification of insulin 2;
- chromatographic purification of insulin 3; - preparation of crystalline insulin.

The recombinant plasmid DNA contains an artificial gene encoding a hybrid polypeptide consisting of one 1g G-binding domain of protein *A* from *Staphylococcus aureus, the* peptide linker His_6GlySerArg and human proinsulin. The plasmid (molecular mass ZD MDa 5051 bp) contains as a genetic marker the ß-lactamase gene, which determines the resistance of bacterial cells transformed with it to ampicillin. The expression of the recombinant protein by the strain-producer is induced by *isopropyl-ß-D-thiogalactoside* (due to the presence of a *tac promoter* in the plasmid) and reaches 30% of the total cellular protein.

Human recombinant preproinsulin producer cells are separated, disintegrated at high pressure to collect inclusion bodies, and the resulting inclusion bodies are repeatedly washed of water-soluble proteins and other cellular components. In one cycle, 15-25 kg of wet inclusion bodies containing more than 13% recombinant protein are obtained. The obtained inclusion bodies are solubilised in a buffer solution containing urea and dithiothreitol (DTT) to restore intra- and intermolecular disulfide bonds in the recombinant protein. The reconstituted recombinant protein is then exposed to air oxygen at high dilution, whereby the protein is oxidised to form disulfide bonds corresponding to the native conformation of insulin (refolding). The recombinant protein is renatured.

Renatured recombinant protein is purified by ion-exchange chromatography and subjected to enzymatic cleavage with trip-blue and carboxypeptidaseB in mass ratio 4000 :2 : 1. After precipitation and pretreatment, 86-88 % purity insulin-serum is obtained. At the first stage of purification by hydrophobic chromatography the insulin content reaches 96 %, at the second stage of purification by ion-exchange chromatography the final liberation from immunogenic impurities is carried out, the final gel-filtration takes place under sterile conditions and allows to bring the quality of the preparation up to pharmacopoeial. At the end of the process, crystalline cyc-insulin is obtained, which is dried to the required degree of humidity. The product yield is at least 100 g cl thousand litres of culture fluid of the producing strain.

Benefits of the process:

- One-step purification of renatured recombinant protein;
- co-enzymatic cleavage of recombinant protein catrypsinomycarboxypeptidaseB;
- Effective use of low-pressure chromatography for insulin purification.

The quality requirements for human insulin substances are set out in the relevant sections of the American Pharmacopoeia (USP) and the European Pharmacopoeia (EP). Insulin must be characterised qualitatively. The first indicator is determined by high-performance liquid chromatography (HPLC) in comparison with the standard, by comparing the chromatographic profiles of insulin and the standard insulin cleaved into 4 fragments by a specific enzyme (peptide map method), and by establishing bioidentity (*in vivo* in rabbits by comparative blood glucose content). Quantification is also performed by reverse-phase HPLC. The amount of hormone is usually estimated in international units (ME), with 1 ME (IU) = 0.0347 mg of insulin. Thyroidin - thyroid preparation

This hormone preparation is obtained from the dried defatted thyroid glands of slaughtered cattle. Thyroid glands are extracted immediately after slaughter from normally developed and healthy animals in slaughterhouses or meat processing plants. They are frozen at -8...-12 °C and delivered in freezers for processing. Before processing, the selected glands are thawed, washed in water, cleaned of fat, connective tissue, muscle, large vessels, etc. Then the glands are crushed and the resulting porridge is spread on flat enamelled plates and dried in a vacuum-drying cabinet at temperatures above 40 °C, after which the material is degreased in a Soxhlet apparatus with organic solvents with a low boiling point, well extracting fats. Residues of organic solvents are removed from the raw material by drying in vacuum dryers at temperatures not exceeding 40 °C. The dry defatted mass is ground in porcelain ball mills. The preparation is standardised according to the content of organically bound iodine, which should be 0.17-0.23 %. If necessary, the preparation is diluted with milk sugar. The action of this preparation is due to the presence of thyroxine and triiodothyronine. Chemically, thyroxine differs from triiodothyronine by the presence of an additional iodine atom. Thyroidine is prescribed in case of insufficient thyroid function.

19.2. Pituitary preparations

Such preparations as corticotropin for injection, zinccorticotropin suspension, thyrotropin, adiposin are obtained from the anterior lobe of the pituitary gland of slaughter animals for medical use. **Corticotropin.** Adrenocorticotropic hormone (AChE) is produced in the basal cells of the anterior lobe of the pituitary gland. It is a polypeptide hormone consisting of 39 amino acids. Its activity is determined biologically and expressed in units of action. Corticotropin is a physiological stimulant of the adrenal cortex.

A widespread method of industrial production of hormones from pituitary gland was developed at the All-Russian Research Institute of Technology of Blood Substitutes and Hormonal Preparations in 1987. It consists in complex processing of raw materials, when separate hormones are sequentially isolated from the anterior-pituitary ligament.

Freshly frozen anterior pituitary lobes are minced and extracted with acidified acetone (1% HC1 solution in 90% acetone). The acidic water-acetone extract is centrifuged. The filter contains ACTH and lactogenic hormone precipitated by acetone. The concentration of acetone in the mixture reaches 92 %. This mixture is retained in the cold at -2...-5 ^{O}C for 10-12 hours. An acidic acetonated precipitate is obtained, which is separated, washed on a Nutch filter with cooled 98 % acetone and dried in air. The acidic acetonated powder is dissolved in water acidified with acetic acid and ammonia solution is gradually added up to pH = 5.0. Lactogenic hormone precipitates at the isoelectric point. The precipitate is separated by centrifugation and used to prepare a lactin preparation. After separation of lactogenic hormone to the residue added ammonium-acetate buffer solution (pH = 5.0) "passed through a column filled with cationite KM-Sephadex K-25. After completion of corticotropin sorption on the ion-exchange resin, it is desorbed with ethanol solution of ammonium-acetate buffer. The adrenocorticotropic hormone is precipitated from the eluent with ethanol. The precipitate is separated by centrifugation, washed with ethanol, acetone and air-dried. The preparation is standardised.

Follicle-stimulating, luteinising and thyroid hormones are further produced from the residues of the original raw materials.

A number of techniques have been developed for similar stepwise release of luteinising, follicle-stimulating and thyroid hormones from tissues following the release of growth hormone, but from the human pituitary gland.

19.3. Growth hormone

Growth hormone was discovered in 1920 and obtained in crystalline form from the pituitary gland of animals in 1944 by scientists D. Lai-emiSevans.

In 1956, human somatotropin was isolated, and in 1958, an endocrinologist at the New England Medical Centre in Boston M. Raben first administered it to a child who was not growing because his body was not producing the hormone at all. The treatment worked, and the baby sprouted.

At that time, the only source of hormone was human brain - cadaver brain - as animal hormone is ineffective for humans.

Most of the cadaver material came from Africa. The hormone was extracted from the pituitary gland, and because it is destroyed by heat'it was pasteurised rather than sternlled in pharmaceutical factories. In the 1980s, three children receiving growth hormone developed a rare viral disease called Creutzfeldt-Jacob disease (CJD). It is characterised by progressive dementia and loss of muscle control. Within about 5 years, the affected person dies.

After the discovery of a disease in udetes treated with growth hormone, the distribution of the drug was stopped. In 1991. In 1991, seven children in the United States developed IBD, and there were 50 cases of the disease associated with growth hormone injections worldwide. And the number of patients could increase, because the disease is caused by an infectious agent, which may not give any symptoms for a long time (up to 15 years). Since cadaver brains as a source of hormone had to be abandoned, the problem of obtaining synthetic hormone arose. Growth hormone is the largest protein produced by the pituitary gland and consists of 191 amino acids.

In 1985, *Genenteçh* created the second drug in history using recombinant DNA. DNA - somatotropin. The genetically engineered growth hormone differed from its human counterpart by a single amino acid. And although this small discrepancy had no effect on the drug's effectiveness in the human body, it opened the door for competitors. The following year, Indianapolis-based pharmaceutical company *Eli Lilly* created a new 191-amino acid growth hormone (Humatotrope) that was 100 per cent identical (physically, chemically and biologically) to that produced by the human pituitary gland. Currently, due to the monopoly of these companies, treatment with the drug costs between 14,000 and 30,000 dollars a year, which creates economic difficulties.

After the hormone is secreted by the gland, it enters the bloodstream where it immediately binds to albumin and other specific carriers, only 5% of the secreted hormone remains in the blood in an unbound state. With the bloodstream, the hormone is delivered to and contacts all organs and tissues. However, its specific action begins after binding to receptors sensitive only to this hormone. It is known that receptors for hormones with peptide, protein and amino acid structures are located on the cell surface, while receptors for hormones derivatives of fatty acids and steroids are located inside the cell (in the cytoplasm or on the membrane of the nucleus). Specific selective absorption of hormonal signal is explained by the highest affinity and sensitivity of receptors and hormones. Hormone deficiency in the body can occur for various reasons and these must always be established. For example, the lack of insulin may be from the fact that it is little produced in the body, it is quickly broken down by insulinase, enters into new compounds; in addition, the structure of efferent systems with iim does not interact. Each of these causes may be due to different factors. Thus, poor insulin production may be based on: a) exhaustion of producing cells; b) suppression of the function of producing cells by alloxan, dehydroascorbic acid, ninhydrin, etc.; c) disruption of some parts of insulin synthesis. It is clear from this that hormonal preparations are effective only under certain conditions.

In accordance with the specificity of pharmacodynamics of hormonal preparations, they are most often used to compensate for the lack of hormone in the body (in case of hormone excess, pharmacological agents counteracting the hormone or reducing the function of the hormone-producing gland are usually used), for specific pharmacotherapy and pharmacostimulation in diseases not related to the lack of hormone in the body.

The use of hormonal, steroidal drugs in sport is widely known. Their detection is no longer of primary importance and control laboratories are required to detect small quantities of designer drugs. Hormonal contraceptives are actively used.

Hormonal preparations are used in both medicine and animal husbandry. Techniques are being developed to raise pigs using regular injections of insulin to accelerate weight gain by administering follicle-stimulating hormone.

19.4. Production of industrially important steroids

Steroids include a large group of biologically important compounds, including sex hormones, cardiac glycosides, bile acids, vitamins, alkaloids, and plant growth regulators. The steroids are based on the perhydrocyclopentanophe-nanthrene skeleton

Biotransformation processes are reactions of transformation of initial organic compounds (precursors) into the target product by means of cells of living organisms or enzymes isolated from them. The ability of microbial cells to perform highly specific biotransformation is used in the production of steroids. Utilising the absolute stereospecificity and substrate specificity of cellular enzymes has enabled a multitude of chemical reactions to be set up for structural rearrangements of steroids. As a result, new compounds with better pharmacological properties were obtained. The biotransformation of steroids usually involves selective action on one of the positions of the steroid skeleton.

The significance of the developed microbial transformation is determined by the fact that the processes of hydroxylation of progesterone and its prons are the basis for the industrial synthesis of many valuable products: anti-inflammatory and antitumour drugs, tranquilizers, anaesthetics, sex hormones, etc. The processes of hydroxylation of progesterone and its prons underlie the industrial synthesis of many valuable products.

In addition to biotransformation, steroid hormones can also be produced by cell cultures. For example, cell culture of *Dioscorea deltoidea* root cells produces phytoste-rnn, diosgenin and its glycoside derivatives (saponins). It is significant that the ability to supersynthesis of furostanol glycosides of a number of Dioscorea strains, and the DM-OH strain, was stably maintained for 27 years. Thus, *in vitro* cultivation of plant cells represents a new solution to the problem of industrial production of secondary metabolites.

In the future, the use of immobilised cells for steroid hormone production, the use of an optimal combination of biological and chemical transformations, as well as the improvement of purification technology of the obtained compounds are predicted.

Chapter 20

BIOTECHNOLOGY

ENZYME PRODUCTION

20.1 Sources of enzyme preparations

Enzymes are found in virtually all plants, animals and microorganisms. Since the process of enzyme biosynthesis in the organism is connected with the provision of cell metabolism and the amount of synthesised enzymes is strictly determined by the vital requirement of the organism, such objects cannot serve as a source of enzyme preparations. Only microorganisms, some plants or individual organs of plants and animals capable of accumulating significant amounts of enzymes are suitable for this purpose.

Plant raw materials. The source of enzymes can be sprouted grain of various cereals (malt). It can either be used directly as a technical enzyme preparation or serve as a starting material for the preparation of purified enzyme preparations. In tropical and subtropical countries, latex of melon tree, latex of plants belonging to the ficus species, e.g. leaves and shoots of figs, juice of green mass of pineapple, etc., are used as raw materials for industrial production of proteinases.

Animal organs and tissues. All meat processing plants collect raw materials containing enzymes, preserve them and use them to produce enzyme preparations. Such raw materials are pancreas, mucous membranes of stomachs and small intestines of pigs, rennet of cattle, rennet of dairy calves and lambs, testes of sexually mature animals. The pancreas contains a large number of diverse enzymes: chymotrypsin, collagenase, elastase, trypsin, amylase, lipase and others. The mucous membrane of pig stomachs and cattle rennet is a source of pepsin and lipase. The rennet (rennet enzyme) is obtained from the rennet of dairy calves and lambs. The testes of sexually mature cattle contain the enzyme hyaluronidase.

Microorganisms. Under specially created conditions, microorganisms are capable of synthesising a great variety of enzymes. They are unpretentious to the composition of the nutrient medium, easily switch from the synthesis of one enzyme to another and have a relatively short growth cycle (16-100 h). Both natural strains of microorganisms isolated from natural objects and mutant strains are used for industrial production of enzyme preparations. Various microorganisms can be enzyme producers: bacteria, fungi, yeasts, actinomycetes. Microorganisms can synthesise simultaneously a whole complex of enzymes, but there are some (especially among mutant strains) that are monoenzymatic and produce large quantities of only one enzyme.

Classification and nomenclature of enzymes and enzyme preparations. According to the modern classification, all enzymes are divided into the following main classes: (1) oxidoreductases; (2) transferases; (3) hydrolases; (4) lyases; (5) isomerases; (6) ligases (synthetases). Most industrially important enzymes, the demand for which is determined by tens of thousands of tonnes, belong to the third class, hydrolases. Usually, preparations produced by companies of different countries are complex, containing, in addition to the main enzyme, a significant number of related enzymes and proteins. For this reason, in enzyme technology, preparations are more often classified according to the main component in the mixture of enzymes present in a given preparation: amylolytic, proteolytic, lipolytic, etc.

In our country, there is a certain system for naming enzyme preparations, which takes into account the main enzyme, the source of production and the degree of purification. The name of each preparation includes the abbreviated name of the main enzyme, the species name of the producer and the preparation, and the suffix -pn. For example, amylolytic preparations derived from *Aspergillus oryzae* and *Bacillus subtilis* are called amilorhizin and amylosubtilin, respectively. An index is then given to indicate the method of production and the degree of purification of the enzyme from ballast substances. In the case of deep cultivation the letter "G" is placed after the name, and in the case of surface cultivation - "P". If it is a crude culture of a producer, the letter "X" follows. Between the letters "P", "G" and "x" there may be a number indicating the degree of purity of the preparation. Index2 denotes liquid crude concentrate of the initial culture; 3 - dry enzyme preparation obtained by drying (spraying) of crude enzyme solution (extract from surface culture or culture fluid); 10 - dry preparations obtained by precipitation of enzymes with organic solvents or by salting. Indexes 15, 18, 20 denote preparations partially freed not only from ballast substances but also from accompanying enzymes. . The nomenclature of preparations with indexes higher than 20 is not used, because in these cases we are talking about highly purified and even homogeneous enzyme preparations, which are named according to the classical nomenclature and classification of enzymes.

Characterisation of the activity of enzyme preparations. Enzymes are substances of protein nature, therefore it is impossible to determine their quantity in a mixture with other proteins. The presence of some enzyme in a given preparation can be established by the results of the reaction catalysed by the enzyme, i.e. by the amount of formed reaction products or reduction of the initial substrate. In quantitative terms, the activity of an enzyme is conventionally determined by the initial rate of the enzymatic reaction.The initial rate depends on many factors, the most important of which are temperature, substrate concentration, pH of the reaction mixture and time from the start of the reaction. For this reason, at the suggestion of the Enzyme Commission of the International Biochemical Union, rules were adopted for the determination of drug activities and their expression in activity units.

Standard unit of activity. This value for any enzyme denotes the amount that catalyses the conversion of 1 μmol of substrate in 1 min under given regulated conditions. In Russian and German this unit is denoted by the letter E, in English, French, Italian and Spanish - U. Often the amount of substrate cannot be expressed in terms of micromoles because the mass of the molecule is not known, e.g. when acting on protein, starch, pectin, cellulose. In these cases, the microequivalent of the groups affected by the reaction is determined. Thus, in protein hydrolysis, the number of free carboxyl or amine groups formed, i.e., cleaved peptide bonds, rather than the number of hydrolysed molecules is taken into account; in starch and polysaccharide hydrolysis, the number of hydrolysed glucosidic bonds is taken into account, etc

The Enzyme Commission recommended the following conditions for determining enzyme activity: substrate concentration, enzyme concentration and pH are chosen to be optimal for the enzyme, temperature is 30 °C; activity is determined from the initial reaction rate when the substrate concentration is sufficient to saturate the enzyme and corresponds to zero-order reaction kinetics.

If the amount of substrate reacted is very small or large, it is acceptable to express results in milliunits (MU and MU) and kilounits (kEic).

Activity of enzyme preparations. The enzyme content in a given preparation is conventionally expressed in standard units of enzyme activity per 1 ml of enzyme solution or 1 g of preparation. The activity of an enzyme preparation is expressed in micromoles of substrate reacted in the presence of 1 ml of enzyme solution or 1 g of preparation under given conditions for 1 min The number of micromoles is equal to the number of standard units. If the enzyme is homogeneous, its specific activity can be expressed in standard units per 1 mg of enzyme. If, however, the preparation contains ballast in the form of inactive protein, its specific activity is expressed in standard units per 1 mg of protein in the enzyme preparation. Molecular activity is the number of millimoles of substrate or equivalents of affected groups reacted within 1 min with 1 mmol of enzyme at optimum substrate concentrations, or the number of standard units contained in 1 mmol of enzyme.

If an enzyme contains a characteristic crostatic group or several catalytic centres that can be measured, its activity can be expressed in terms of the catalytic centre activity. It will correspond to the molecular activity if the enzyme molecule has one active centre; if the number of catalytic centres is *n,* the activity of one centre will be *n* times less than the molecular activity.

Activity of a conditional preparation. In enzyme technology, in addition to the generally accepted notion of enzyme preparations activity, it is customary to use the notion of activity of a conditional enzyme preparation. This is necessary to evaluate the work of the enterprise, to compare it with similar plants, i.e. to compare the indicators for all types of products. For the purposes of such recalculation, it is assumed that the enterprise produces marketable products in the form of a standard preparation with a precisely defined activity, measured by the main enzyme in standard units in the preparation per unit mass of the preparation. The activity of the main enzyme in such a standardised conditional preparation is established by regulations and is referred to as the activity of the standard preparation.

For 1 ust of enzyme preparation 1 tonne of preparation with standard activity is taken. The formula is used to convert the produced marketable products into conditional tonnes

$$Qusl - QTOV \ \mathrm{F} / \mathrm{usl}'$$

where Qu_{SL} - quantity of conditioned drug, tonnes; Q_{tov} - quantity of commercial drug, tonnes; A_f - actual activity of commercial drug, units/g; $A_{(usl)}$) - activity of conditioned drug, units/g.

20.2 Applications and sources of enzymes

The production of enzyme preparations occupies one of the leading positions in modern biotechnology. The volume of their production is constantly increasing and the scope of their application is expanding. Enzymes are highly active non-toxic biocatalysts of protein origin. Their advantage over chemical catalysts is their action at normal pressure, temperature 20...70 C, pH from 4 to 9. They have high substrate specificity, which makes it possible to target only certain compounds in a complex mixture of substrates.

According to the accepted classification "nomenclature, about 2 thousand enzymes have been identified. About 250 items are commercially produced, and 99% of the total sales of enzyme preparations are only 18 enzymes.

Proteinases, widely used in synthetic detergents, and amylases for starch processing occupy the largest share among the produced preparations. These two types of preparations account for 60% of the total output of enzyme preparations abroad.

Other major enzyme consuming industries are[are;]

- wine and juice production (10%);

- alcohol production (8%);
- s "R? Dejaye(5%);

- baking (5%);

. brewing(b%);

. other industries" %).

In addition, in Russia, enzymes are being introduced into fodder production.

Highly purified enzyme preparations occupy a special place in the total volume of enzyme production. Their share in the total volume is very small, as the technology is complex and requires high material costs and time. These preparations are very important for medicine, analytical purposes and scientific research.

Enzymes are formed in all living things, but only those natural objects with an enzyme content of at least 1 % of the enzyme used are used for their isolation.

Sources of enzymes can be:

. germinated grain of various cereals (malt) - for obtaining proteinases, ficus latex, melon tree - for obtaining proteinases;

. individual tissues and organs of animals (pancreas, mucous membranes of stomachs and small intestines, cheese rennet of cattle);

. microorganisms. Under specific conditions, microorganisms are capable of synthesising a huge number of enzymes. They easily switch from the synthesis of one enzyme to another and have a short growth cycle (from 16 to 100 h). For industrial production of enzymes, both natural strains and those obtained by mutagenesis, selection and induction of biosynthesis are used.

Enzymes can be synthesised by bacteria, fungi, yeasts, actinomycetes; microorganisms can be mono- or poly-enzymes.

20.3. Selection of strain and cultivation conditions

The feasibility of using microorganisms for enzyme production is as follows:

1) genetic manipulations can increase the level of catabolic enzymes by a thousand or more times and the level of biosynthetic enzymes by several hundred times;

2) It is energetically justified to grow microbial cells on a large scale due to the use of inexpensive media and rapid growth.of microorganisms;

3) the great variety of reactions to which microorganisms are capable, which is especially true for secondary metabolism;

-1) microorganisms serve as a source of not only such enzymes found in animals and plants, but also a number of unique enzymes not found elsewhere (e.g. cellulose, tannase, hydrogenase, etc.). Among microorganisms there are species that develop at extremely high temperatures

(B7 -C), and therefore it is potentially possible to create thermostable strains; 5) the ability of microorganisms to adapt to different environmental conditions, which allows the culture to be transferred to production facilities where it grows on cheap substrates.

The industrial production and application of enzymes is based on two important factors; firstly, enzymes are formed in living cells; secondly, they can exert their specific action in an independent manner in living cells.

The initial isolation of strains is based on the assumption that microorganisms are adapted to the cotyledonisation of substrate that is abundant in their habitats and therefore produce enzymes that react with this substrate. For this reason, cellulase and lignolytic enzyme producers are isolated from forest soils, pectinase producers from fruits and plants.

Enzymes from closely related strains have similar properties, while those from distantly related strains can differ greatly. The first step in enzyme production is to select the organism that produces the desired enzyme in the highest quantity. The following general requirements for the producer are taken into account:

- the obligatory formation of extracellular enzymes because they're easier to isolate;

. high enzyme yield in a short time;

- easy purification of the enzyme from the culture fluid;
- strains must not produce antibiotics, toxic substances or be related to toxin-producing strains.

To induce enzyme supersynthesis, a number of methods are used that involve altering growth conditions and leading to changes in the genetics of the organism.

20.4. Technology of cultivation of enzyme-producing microorganisms and isolation of enzymes

The technological process can be broken down into three stages: . receiving seed material;

- development of a production culture by surface or deep cultivation methods;
- isolation of technical or purified enzyme preparations from the finished production culture.

The surface method consists of culturing microorganisms on the surface of moistened sterilised bran spread in cuvettes. Incubation is carried out in a special thermo-metatised room with constant control of temperature, humidity and air supply.

The deep method is more economical. For its implementation, stainless steel fermenters are used, equipped with devices for mixing and supplying sterile air to the liquid nutrient medium.

The most advanced is the flow method, which provides continuous supply of nutrient medium and seed materials to the fermenter, continuous sampling of life products and microbial mass. The advantage of this method is the possibility to maintain the growth of microorganisms culture in automatic mode for a long time (up to 200 days).

The isolation and purification of enzyme from microbial culture. - is a rather labour-intensive and expensive procedure, so if the enzyme can be used in the form of a crude preparation, it is not purified. In industry, commercial preparations with a purity of only 0.1 % (i.e. 99.9 % impurities) are widely used. Such industries include the alcohol, leather, textile, agriculture and household chemical industries.

The crude preparations are obtained by gentle drying of the microbial culture together with the residual nutrient medium. Such preparations are obtained either from the culture extract of a surface-grown producer or from the culture liquid filtrate of a deep-grown producer.

Most food, research and medical applications require purified enzyme preparations. The source of enzyme extraction can be microbial biomass, extract or culture fluid filtrate. Microbial biomass must be thoroughly pulverised to the point of destruction of subcellular structures: lysosomes, mitochondria, nuclei, etc.

The preparations are then purified by precipitation with organic solvents, salts followed by dialysis, affinity chromatography, re-precipitation, gel filtration, sorption, crystallisation, etc.

Since enzymes are of protein nature, special importance is attached to the observance of regimes that preserve their activity while inactivating associated ballast proteins. Purified enzymes are stored at low temperature (down to -80 °C). To stabilise enzymes, coenzymes and substrates are added to their preparations.

Enzymes are widely used as biological catalysts in various fields of human activity. Their sources can be animals, plants and microorganisms. To date, more than two thousand enzymes have been identified, and several hundred of them have been obtained as individual substances.

Microorganisms as enzyme producers are of particular interest because their metabolism, and consequently the work of enzyme systems, is carried out at a very high intensity.

If we characterise the total intensity of enzyme reactions on absorption of molecular oxygen in the form of the coefficient *QO* (number of millimetres of oxygen) consumed by 1 mg of dry biomass in one hour, then for liver cells it is equal to 2-5; kidney - 1020, yeast cell - 50-110; for bacteria of *the* genus *Acetobacter* - 1,8 thousand, *for Azotobacter - 2* thousand.

In addition to the high intensity of metabolism, it should be borne in mind the very high rate of biomass growth of microorganisms This allows for short periods of time (sometimes 24-72 h) to obtain such quantities of raw materials for the isolation of enzymes, which can not be compared with what is given by plants and animals.

An important property of many microorganisms is their ability to grow on various cheap substrates, including those that have no nutritional value (cellulose, petroleum hydrocarbons, methane, methanol, etc.). Some substrates used by microorganisms are toxic to humans and animals.

The content of individual enzymes in the cells of microorganisms can be very high. For example, the amount of ribulose bisphosphate carboxylase in phototrophic bacteria sometimes reaches 40-60% of all soluble proteins.

Many microorganisms produce enzymes that are released in large quantities into the culture medium. These enzymes mainly belong to the hydrolases that break down proteins, starch, cellulose, fats and other water insoluble substances.

The isolation of enzyme preparations from microbial cells and especially from the culture medium after their cultivation is easier and more economical than from plant and animal tissues.

A number of enzymes are found only in microorganisms. Such enzymes include: tannase, which cleaves digallate to gallic acid, racemases of many amino acids, keratinases, which hydrolyse sulphur-containing proteins - keratins, which are part of hair, feathers, horns and hooves. Some microorganisms possess specific decarboxylases of amino acids, form penicillinase, which cleaves penicillin of dopenicillin acid and water

Such an enzyme as nitrogenase, which is involved in the formation of ammonia from molecular nitrogen, has been found only in bacteria capable of H_2 fixation.

Some bacteria are able to oxidise inorganic substrates: ammonia, nitrite, sulphide and other compounds, as well as divalent iron, which is due to the presence of special enzymes. A number of bacteria and algae synthesise hydrogenases that catalyse the oxidation and formation of molecular hydrogen.

A significant number of bacteria are capable of synthesising enzymes that allow them to use methane, methanol, methylated amines, carbon monoxide and other one-carbon compounds as substrates for.growth.

The ability of enzymes produced by microorganisms to break down plastic components, pesticides and other toxic compounds makes it possible to cleanse the environment of a number of pollutants.
The ability of microorganisms to develop in extreme environments, i.e. at low and high temperatures, in the absence of molecular oxygen in acidic or alkaline environments, at high salt concentrations, is often determined by the nature of their enzymes.
The psychrophilic yeast *Candida gelida* has a growth optimum at 15 °C; when heated to 35 °C, it dies: only pyruvate carboxylase is irreversibly inactivated. In the obligate-psychrophilic strain *Micrococcus cryophilus*, short-term heating to 25 °C changes the secondary and tertiary structure of glutamine and proline-TP-NK synthetases. As a result, cells lose the ability to link tRNA to the appropriate amino acid. Some pseudomonads show active growth at 3...4 °C and their enzymes are active at -5...-10 °C. An exceptional position among living creatures is occupied by thermophilic bacteria, which have optimum growth at 60...80 °C and above. The possibility of development of these bacteria at high temperatures is conditioned by thermostability of proteins. For example, the extreme thermophile TTzertnusaquaticus possesses enolase with optimum action at 90;C
The fungus *Malbrancheapulchella* var. *sulfúrica*, which can grow at 45 °C, secretes a thermostable proteinase into the medium that remains active at 73 °C.
Some mesophilic microorganisms also produce enzymes with high thermostability, which exhibit maximum activity at a higher temperature than that required for growth.
The life form of halophilic microorganisms is unusual; their growth is observed even in saturated Nad solution. Enzymes of such microorganisms require high concentrations of salts for their maximum activity. The relationship of microbial enzymes to the acidity and alkalinity of the medium is also unique in many cases.
The amylolytic activity of *Asperg nger* is provided by two amylases, of which one is stable at very low pH values.
In contrast, some amylases require sharply alkaline conditions for activity. For example, *Bacillus* sp. amylase showed maximum activity at pH 10.5. Lime bacteria that grow in an alkaline environment (pH 9.0 and above) and produce enzymes with a high pH.
Of great interest as enzyme producers are anaerobic microorganisms, many of which convert amino acids, purines, pyrimidines and other substrates by pathways different from those known for aerobic microorganisms, animals

and plants.

In recent years, it has been established that among strict anaerobes there are bacteria capable of growing under chemolithoautotrophic conditions. In this case, they have systems of carbon dioxide assimilation that are not typical of other organisms, which is due to the presence of special enzymes. Some methane-, acetate- and sulfate-reducing bacteria belong to such autotrophs.

The attention paid to anaerobes is explained by the possibility of obtaining ethanol, butanol, methane, acetate and other useful products from their activity during the processing of plant residues and other cheap raw materials. Cultivation of anaerobes eliminates the need for aeration and mixing of the nutrient medium; the possibility of infection caused by insufficient air sterility is reduced. In addition, the use of anaerobic microorganisms allows the process to be carried out in large containers with high media volume.

A property of many enzymes of microbial origin is their inducibility. For example, the synthesis of ß-galactosidase in *Eschenchia coli* is induced by lactose and begins less than 3 min after its introduction, and when the inducer is removed, the synthesis stops as quickly. For this reason, by changing the cultivation conditions, it is possible to achieve the formation of large amounts of many practical enzymes by microorganisms

It is known that enzymes catalysing the same reaction, but obtained from different organisms, may differ significantly in their properties. An example of this is glutaminase, which carries out deamidation of glutamine to form glutamn- new acid. Glutaminases derived from *Azotobacter agilis* cells have an optimum pH = 6.0-7.0; from *Clostridium welchii*, 5.0; from *Mycobacterium tuberculosis*, 8.3. Chlorides inhibit glutaminase of *Pseudomonas* sp. but do not affect its activity *in E coli* and activate glutaminase of *CI. welchii*. Asparaginases of *E. coli* and *Alcaligenes euirophus* differ in sub-strate specificity.

In summary, it should be emphasised that microorganisms have a very high activity of enzymatic reactions and are able to carry out many processes absent in macroorganisms due to the presence of specific enzymes.

Enzymes of similar function in micro- and macroorganisms can be different in properties, and *in* microorganisms they sometimes need special conditions to manifest their activity, so the study of enzymes of different microorganisms is a very important task.

Obtaining active producers. Mutants in most cases are auxotrophic in a number of compounds, as they have undergone certain metabolic disturbances causing hypertrophy of some cell functions. Usually, active strains identified from natural sources are subjected to the action of mutagens several times, i.e. stepwise selection is carried out. As a result, highly productive strains are obtained. The combined effect of mutagens of chemical and physical nature is often effective. For example, the use of ethylenimine and ultraviolet radiation in combination with stepwise selection made it possible to obtain very active strains of *Asp. awamori* used as producers of amylolytic, proteolytic and other enzyme complexes. Selection of industrially valuable strains is also being carried out at other enterprises.

An indispensable condition for a properly organised industrial process are such measures for the preservation of microorganisms-producers. There are a number of methods for preserving production-valuable strains, ensuring their high biochemical activity. In museums of living cultures at factory laboratories, cultures are periodically crossed. However, reseeding at short intervals of time can reduce the activity of microorganisms, so after the development of culture on a dense medium, it is poured sterile vaseline oil. In many cases, the best method of storage is to lyophilise cultures.

Nutrient media for cultivation of microorganisms. Nutrient media by their purpose are divided into three main groups: for the maintenance of stammas-producers in the laboratory and in the museum; for obtaining inoculum; used in large volumes in the main production process. If the first group of media is selected on the basis of physiological needs of microorganisms, the last two should meet the requirements of the industry, i.e. should be cheap enough. The composition of media for the accumulation of a particular enzyme may differ significantly from the composition of media used for the isolation and maintenance of producer cultures.

As noted above, some enzymes of microorganisms are inducible and require the presence of an inducer in the medium for their synthesis, and the inducer may not always be the substrate on which the enzyme acts. Thus, for example, if for the synthesis of lipase *of Aps. niger lipase* is favoured by the presence of vegetable oil in the medium and nucleic acids in the medium for the synthesis of ribonuclease, while for the increased synthesis of α-amylase *by Vas. potymyxa it is not enough to have the presence of the substrate. potymyxa*, the presence of soluble starch in the medium is not sufficient, and the addition of cazenne hydrolysate is necessary. When organic nitrogen compounds were replaced by inorganic ones, the enzyme was not synthesised. In order to reduce the cost of enzyme production, the cheapest sources of raw materials suitable for cultivation of microorganisms-producers are selected.

In industrial production of hydrolases, various wastes containing specific substrates are usually used. Thus, in the production of pectolytic enzymes, pectin-containing raw materials, such as beetroot cake, are used; for cellulolytic enzymes, straw, bran and other cellulose-containing substrates are used. However, the use of media with such insoluble substrates in industry is not always technologically feasible, and also complicates the directional fermentation during deep cultivation of producers.

The use of media with soluble carbon compounds makes it possible to eliminate these disadvantages, shorten the fermentation time, and conduct the process at the same level. Milk whey, a cheap by-product of cheese, cottage cheese, and casein production, is recognised as a promising raw material for the production of some microbial exoenzymes (pectinase, cellulase, etc.). The use of milk whey as a nutrient medium also allows to realise a continuous method of obtaining a number of exoenzymes.

A cheap nutrient medium with wheat so- loma vvidechekimuki was selected for cellulase production using a thermotolerant strain of *Asp. terreus.*

Biomass of other microorganisms subjected to hydrolysis or extraction can be used in the preparation of media for industrial cultivation of enzyme producers

A method of extracting the biomass of pectinase-producing fungi and the subsequent introduction of extracts into the nutrient medium has been developed, which allows to fully utilise all the biomass obtained, reducing the amount of beet pulp extract by 9% and intensifying the biosynthesis of pectolntntnickhfermeitov. 2.4 times.

A closed technological scheme for the production of pectolytic enzymes has also been developed, which makes it possible to polyostyutilise microbiological production wastes by returning them to the technological process.

The search for new nutrient media that meet the requirements of modern enzyme production and can replace the initial raw materials (feed and food products) is an urgent problem of biotechnology.

20.5. Extraction and stabilisation of enzymes

Methods of isolation and purification of microbial enzymes are different and are determined depending on the localisation of the enzyme (in cells or culture medium) and the purpose of application. Unpurified enzyme preparations are obtained by drying and crushing of fungal mycelium together with solid substrate (bran, cake, etc.) or drying of the producer together with the culture liquid on a spray dryerDried preparations are crushed into powder for further use. The specific enzyme activity of such preparations is low, but they are cheaper and sufficiently stable during storage. This method is used to produce amnlase-prote- ases, cellulases for agriculture and some industries.

Syrups are obtained by thickening the culture liquid separated from the biomass of the microorganism containing the enzyme.

In laboratory practice and industry there are various methods of concentration, but only a few of them can be used to thicken enzyme solutions.

In the industrial production of enzymes, along with activity, their thermal stability, pH and stability are important parameters. It should also be taken into account that an increase in the concentration of crude enzyme solutions may be accompanied by an increase in the content of inhibitory impurities up to critical with simultaneous decomposition and removal of stabilising products. But usually decrease of water content during thickening leads to decrease of speed of chemical and biochemical processes.

There are several methods of concentration: 1) without phase change (membrane concentration); 2) with phase change (evaporation, evaporation). Concentration with membranes is based on the fact that the composition of the liquid can be changed by passing it through a membrane with selective permeability. The ideal membrane should allow the desired substance to pass quickly through the solution, but should also act as a barrier to all other components.

Another membrane variant is also possible, in which the required substance is retained and all other components are allowed to pass through. In practice, membranes of both types are used.

To thicken thermosensitive liquids, a technique based on the selective crystallisation of water in solution at cooling below O ^{0}C is used. The essence of this method of concentration is that the water of the thickened solution turns into ice crystals, thus increasing the concentration of soluble substances. The ice crystals are removed by centrifugation or pressing. The operation is repeated many times until the required enzyme concentration is obtained. Under such conditions the loss of thermostable components of the solution is completely excluded (no thermal destruction).

The development of new methods of concentrating biological fluids does not diminish the importance of such a well-known technique as vacuum evaporation. The abundance of evaporator designs and the state-of-the-art research in this field offer great opportunities to utilise the evaporation process.

The physical meaning of evaporation of solutions is to separate them into two parts: water removed as vapour and concentrate enriched with soluble substances. In enzyme production, the process of concentrating aqueous solutions under vacuum is widespread. For example, the thickening of aqueous extract from mycelium of *Asp.* oryzae is carried out in this way to obtain the preparation of proteorisin. A large number of enzyme preparations are produced using the vacuum concentration method.

Enzyme preparations are dried in vacuum by spraying or frozen. After drying the preparation in case of non-stability should be mixed with stabiliser or with filler (starch, dextrins, inorganic neutral compounds, talc, etc.). In some .cases, the enzyme is stabilised by the addition of chelating agents to the culture fluid. Thus, the alkaline protease of *Vas. subtìlis* was stabilised by adding sodium tripolyphosphate, sodium nitrilotriacetate and ethylenediaminetetraacetate to the medium.

Enzymes are also used as homogeneous proteins. The preparation of enzymes in pure form is carried out by various methods depending on their properties and application. The usual scheme for the preparation of exoenzymes may be as follows:

I. Culture fluid ↓ *centrifugation* Supernatant liquid ↓ *dialysis* Dialysate ↓ *vacuum evaporation* Concentrate ↓ *ammonium salt precipitation* ↓ *dialysis* Aqueous solution ↓ *gel filtration* Desalinated solution ↓ *evaporation*	II. Enzyme preparation Desalinated solution ↓ *acetone precipitation* Filtrate ↓ *acetone fractionation* Precipitate ↓ Aqueous solution ↓ *gel filtration* Enzyme solution ↓ *vacuum drying* III. Pure enzyme

In industry, a high degree of enzyme purification is not always necessary. The purification must be very high if the enzymes are to be used as therapeutic agents.

Purification and separation of enzymes forming complexes are carried out using ion exchange resins. The ion-exchange method, for example, allows the separation of enzymes of the pectolytic complex of fungi, purification and isolation of pectinesterases, polygalacturonases and pectin traszliminases.

Microencapsulation is sometimes used to use an enzyme as a therapeutic agent. The enzyme, enclosed in special capsules, reaches the organ or tissue where it is to show its activity, there the capsule dissolves and the enzyme enters the tissue.

Currently, the so-called immobilised enzymes, i.e. enzymes firmly bound to various carriers, are increasingly used. In addition to immobilisation of highly purified enzymes, the method of immobilisation of whole cells of microorganisms possessing one or another enzyme activity is very promising. In this case, the enzyme is more stable.

The process of purification of the enzyme from ballast substances is reduced to freeing it from insoluble substances, concomitant soluble substances and other enzymes. The processes for obtaining purified preparations from surface and deep cultures are somewhat different. It is more difficult to obtain high value preparations from surface crops because of the large amount of ballast substances. It is slightly easier to obtain purified preparations from deep cultures, but the extraction from dilute solutions has to be carried out. This is more difficult if the enzyme is intracellular, in which case it is necessary to destroy the microbial cells.

From Table 11, it is obvious that the extract from the surface culture or culture filtrate is the starting material for obtaining enzymes of various purification degrees. At the first stage of isolation, the process waste is the insoluble part of the culture bioshrot containing insoluble inclusions of the medium and the biomass of the producer.

Further, depending on the properties of the extracted enzyme and its accompanying ballast, the scheme of purification and preparation of the enzyme preparation may include various techniques and methods such as concentration, dialysis, precipitation with organic solvents, salts, gel filtration, affinity chromatography, immobilisation,

Table 11

Principal scheme of preparation of purified enzyme preparations from microorganism cultures

Purification stage	Volume, ml	Total protein, mg	Glucosaminylase activity				Amylolytic activity		Transglucosidase activity*
			total, units.	specific, units/mg	yield, %	degree of purification	total, units.	yield, %	

				protein a					
Initial culture fluid	1200	13 600	28 500	2,1	100,0	1,0	9500	100	Glucose, isomaltose, panose.
Biomass separation, concentration, ballast separation	560	11 100	25 600	2,3	90,0	1,1	8500	89,00	
Acetone precipitation, dissolution in water	350	2040	19 800	9,7	69,5	4,6	1050	11,30	
Ultrafiltration	55	1610	18 200	11,3	64,0	5,4	860	9,10	
Chromatography on DEAE-cellulose	555	298	15 000	50,4	52,5	245	30,0	0,35	No
Ultrafiltration	16	250	13 450	54,0	47,2	25,7	27,5	0,29	
Gel filtration through acrylex P-100	100	140	11 400	76,5	40,5	36,5	22,8	0,24	
Desalting, lyophilisation	0,1	92	7150	77,0	25,0	37,0	14,3	0,15	

* Characterised chromatographically by the presence of isosaccharides.

* Expressed in grams.

drying of thermolabile materials, etc. On this basis, we will initially consider the production of technical enzyme preparations, and then, without special emphasis on the surface or deep cultivation method, more complex methods of purification and isolation of enzymes.

20.6. Receipt of crude herment preparations

Unpurified enzyme preparations represent a culture of microorganism together with the residues of nutrient medium, dried under mild conditions to a moisture content of no more than 8-12 %.

The crude enzyme preparation can be obtained from surface or deep culture. The deep culture can be purified from the insoluble part (solid suspension of medium and biomass of the producer)™™ before drying.

Obtaining a dry surface culture. The surface culture of the microorganism has a moisture content of 35 to 58 %. It is a low-stable product that should either be used immediately in production or dried to an equilibrium moisture content (10-12 %). Before drying, the culture discharged from the plant chamber is pulverised to a certain particle size and then dried.

Belt, tunnel, shaft drum, cabinet and vibratory dryers can be used for drying microorganism culture. Typically, direct flow type drum dryers are used. The wet crop enters the dryer simultaneously with a heat carrier having a temperature of 80...85-C. Such a high temperature is acceptable to use because the dried material contains a large amount of moisture, and when it evaporates, the particles of culture soil- ™ are not heated and the activity of enzymes is preserved almost completely. Most drum dryers have a vane or cross-shaped nozzle on the inside of the drum. The drum rotates slowly at a frequency of 3 to 8 min^{-1}. The material to be dried is lifted, poured and moved along the drum with the help of the paddles. There is some differentiation of crop particles by size - larger particles fall almost vertically downwards and are picked up again by the blades, while smaller particles are picked up by the coolant, their trajectory is slightly shifted and they move faster.

The product is dried in this dryer with a uniform moisture content throughout the entire mass. Thus, the product dried in this dryer has uniform moisture content throughout the whole mass. The duration of stay of the dried part in the dryer is 3-7 min, the speed of movement of the supplied coolant is 2-3 m/s, the air temperature at the inlet is 80...85 °C, at the outlet is 60...65 °C, the temperature of the dried material is 40 ^{O}C. Loss of activity in the drying process isZ-10 %.

Another type of dryer is the steam conveyor dryer, which is a sealed belt conveyor. When drying in such units, the loss of activity is greater, but they are very compact and have a high capacity. However, it should be noted that losses can be up to 10-20 %. This is due to the fact that there is no differentiation of culture particles by size, small and large particles are in the dryer for the same time, small particles dry out a lot, enzymes are inactivated faster, large particles are slightly underdried, in them the activity losses are lower. But they may increase due to increased humidity during storage.

A wide variety of dryer designs can be used for surface culture drying, in which the duration of culture residence is reduced to 5-8 min with a product outlet temperature of no more than 40...42 °C, thus minimising loss of reactivity.

The finished dry crop is usually packed on a special packing machine in 25-40kg bags, waterproof bags are sewn up on a sewing machine and sent to the finished product warehouse.

Purification of culture liquid from solid suspensions. Most producers accumulate the main part of their synthesised enzymes in the nutrient medium. When obtaining purified enzyme preparations, the insoluble part of the medium together with the biomass of the producer is separated in filters, centrifuges or separators.

In the microbiological industry the most widely used cellular drum vacuum filter of continuous action with an external filtering surface. These filters have a high degree of mechanisation and allow filtering of various suspensions at a constant speed. Drum vacuum filters are a drum immersed in a container, into which the culture liquid is continuously fed. The surface of the drum is perforated and covered with filter fabric (batiste or synthetic fabric of similar type).

Sometimes, in the presence of difficult to separate sediments with high resistivity in the culture liquid, a soap bed is used as a filtering surface. Sludge removal on these filters is carried out with a special knife. With each revolution of the drum, a part of the soap bed is removed together with the sludge and the filter surface is renewed. The drum rotates slowly at a frequency of 0.13-0.26 rpm^{-1} and passes successively through the filtering, drying, sludge washing, drying and blowing zones. The drum is divided into sections by a fixed distributor head consisting of several chambers, which are connected to the vacuum tanks of the filtrate and wash water respectively, as well as to the compressed air line.

Drum filters are suitable for separating not only the biomass of the producer, but also insoluble suspended solids, which are relatively abundant in the medium (pomace, bran, cake, sprouts, etc.). The disadvantages of filters of this type include their relatively low capacity, bulky (the ratio of specific surface area of filtration to the volume of filtrate is small) and impossibility to provide aseptic conditions.

In the fermentation industry, batch filter presses with manual sludge discharge are less frequently used. These filters can produce clear filtrates, but they operate periodically without regeneration of the filter surface. As the sludge space is limited and the sludge layer reaches a considerable thickness at the end of filtration, the filtering speed drops despite the increased operating pressure. When the sludge space is filled with sludge, the filter press is switched off, disassembled and the filter cloth is washed or changed. The capacity of a filter press is smaller than that of a drum vacuum filter. It is limited by the sludge content in the filtered liquid and the volume of the frame space of the filter press. The filtering process in the frame filter is carried out under pressure of 0.6-0.4 MPa. The culture liquid through the holes in the frame wall enters the inner cavity of the filter element, the suspension is retained on the filtering surfaces, and the filtrate passes through the filtering cloth and flows through the grooves in the plates into the pipeline. Usually the first portions of filtrate are turbid and they are re-filtered.

The disadvantages of the filter press are largely eliminated in the horizontal chamber construction. This filter press consists of a number of horizontal filter plates placed one above the other, between which the filter cloth is stretched. The filter plates are placed between the upper and lower support plates and the filter cloth is stretched on guide rollers.

The filter press operation cycle consists of plate compression, filtering, washing and dewatering of the sludge, plate sliding and sludge unloading at the same time as cloth moving and washing. The operation of the FPAKM filter press is fully automated. These filters have an extensive filtering surface (up to 25 $m^{(2)}$ of filtering surface per 8 m^2 of plant area). The sludge is squeezed under pressure of 0.8-1.5MPa and has humidity not more than 60-70%; relatively low energy consumption (0.8-1 kWh per 1m of filtering surface) is observed. Its specific productivity is 6-8 times higher than that of other filter presses (at concentration of solid phase in suspension 4-7 g/l up to 1 thousand litres/(m($^{2)}$ - h), activity losses do not exceed 4-5 %). FPAKM units are produced with filtering surface area from 2.5 to 50 m^2. Their application for purification of enzyme solutions is very promising. This type of filters is especially recommended for filtration of bacterial culture liquid suspension.

Various designs of sheet filters belong to the pressurised filters. They have in common the presence of flat filter elements with a rigid frame. Sediment can be removed from the filter surface by various methods: compressed air, steam, vibration, centrifugal force.

Some bacterial cultures are filtered at rates below 30 l/(m - h) even with the use of auxiliary filter media. In these cases, centrifugal separators are widely used to separate the biomass and remove suspended solids. In enzyme production, clarifiers of the VSM type are used, which is a container with a drum inside. Inside the drum there are concentric cylinders-inserts. The solution to be purified enters the cylinder with the smallest radius through a spigot, then passes along the installed cylinders, each time changing direction. The clarified liquid is removed from the drum by a pressurised pressure disc and the sediment is thrown to the inner wall of the cn- lindrical inserts under the action of centrifugal force. The capacity of this type of separator can reach 2-5 thousand litres per hour.

Various types and designs of self-unloading separators are used in the production of enzyme preparations. In our country we use separators of ASE-3, ASI, ASE-B types with centrifugal pulsating sludge discharging, which have a capacity of 500, 1500 and 2000 l/h, respectively, with a drum diameter of 600 mm and a plate gap of 0.5 mm. Scientific enterprises use the so-called nozzle separators of the *Alfa Laval* company (Sweden) of the QX and RECX types with the capacity from SZO to 200 m^3/h.

The *Alfa Laval* company manufactures actually deflattening nozzle high-speed separators - so-called bactofuges of the $D^3$187M type (capacity of b m^3/h) and AX-213 (capacity up to 36 m^3/h). The separation factor, for example, on bactofuge AX-213 is 142 thousand, which makes it possible to obtain a fugate, almost completely cleaned from microorganisms and the finest suspension, which is very important.

The digester is mounted on a bed. It has a shaft with friction clutch and brake, worm gear, vertical hollow rotor spindle with feed pump. A drum with a set of conical plates is mounted on the hollow shaft. The upper part of the bed, together with the hood and the liquid removal device, is enclosed in a cooling jacket, which ensures low temperatures during separation - this is extremely important, as higher temperatures during sludge separation can lead to inactivation of enzymes. The slurry enters the rotor from below through the hollow shaft and is distributed over the plates by centrifugal force. The solids are directed towards the walls of the rotor and are continuously discharged through the con- la with a small amount of liquid. The bulk of the liquid is removed from the bactofuge under pressure through the upper paraxial outlet. The clean filtrate is removed continuously.

The microbial cells and other suspended particles with a small amount of liquid are collected under the cover above the rotor and flow down the inlet pipe along the cover to the cyclone where they are deaerated. The contaminated air is directed to the top of the rotor lid where it is again mixed with the microbial cells exiting through the nozzles - a closed cycle is formed. The concentrate of microbial cells, separated from the air, is removed through the bottom nozzle of the cyclone.

This type of separator has great advantages: process accuracy, continuity of suspension loading and filtrate extraction, purification of air from the producer or enzyme, possibility of process operation at low temperatures, etc. It can be safely stated that this type of separators will be successfully used in the enzyme industry in the XXI century. However, it should be noted that the efficiency of biomass separation largely depends not only on the types of apparatuses used, but also on the composition of the medium, the size of separated particles, the amount of insoluble fraction, the physical and chemical characteristics of filter materials, and temperature conditions.

In order to improve the filtration process, the culture liquid is pre-chemically treated. For this purpose, the culture liquid is alkalised to pH8-8.5 and 0.1% solution of calcium chloride is introduced, resulting in the formation of calcium phosphate gel, which contributes to the most complete separation of the sludge at the lowest losses. However, chemical pretreatment does not always give good results, so to increase the efficiency of the process, various diatomaceous earths are often used, such as diatomite and radiolite (Japan), microsil (France), diatomite (Belgium), clargel (UK), etc. The use of these fillers can sharply increase the speed of filtration, but at the same time increase at teriactivostyna this technological stage.

The resulting biomass of the producer together with the insoluble media particles (bioshrot) is sterilised, dried and used as animal feed if necessary. The culture liquid filtrate is unstable, cannot be stored and should be immediately sent for further processing to obtain purified enzyme preparations.

20.7. Extraction of enzymes from surface cultures

All enzymes are water-soluble proteins, so the best extractant for them is water. To extract enzymes from yeast or bacteria, it is necessary to subject their cell walls, which have a high diffusion resistance, to mechanical or autolytic destruction. Mycelial filament envelopes have lower diffusion resistance than bacterial and yeast cell envelopes, so disintegration of the fungal culture is not required.

Enzyme extraction is carried out from both wet and dry surface cultures of fungi. Dry cultures can be stored for a long time without loss of enzyme activity, and more concentrated extracts are obtained from them. Technologically it is more favourable, but during drying of the culture there are losses of activity," therefore extraction should be carried out from the wet culture. During extraction various water-soluble substances are extracted from the culture with unequal speed, their partial fractionation occurs, specific activity of enzymes in the extract increases 3.5-4 times compared to the initial culture as a result of separation of a large part of substances (up to 75 %) with insoluble residue - bnoshrot.

The completeness of enzyme extraction from cultures is influenced by many factors: temperature, pH, process duration, design of extraction apparatus, nature of the extracted enzyme, amount of extract extract extracted per unit mass of culture loaded in the apparatus, etc.

Many other compounds are extracted simultaneously with enzymes, and often the extraction rate of ballast substances is greater than the extraction rate of the target enzyme culture.

Thus, it is more rational to make some losses of the enzyme and to finish the extraction at the optimum value of the ratio of the enzyme activity in the extract to the sum of extracted substances. This question is solved experimentally for each species of producer.

It is practically impossible to influence the extraction process by temperature, as the enzymes are very thermolabile and are inactivated even at 35...40 ^{0}C. In addition, such an increase in temperature leads to an increase in the dry matter content of the extract and a decrease in the specific enzymatic activity per 1g dry matter, increasing the risk of infection of the extracts. On this basis, when carrying out extraction in factory conditions, they try to suppress the development of microflora by maximum reduction of water temperature to 22...25 ^{0}C and the use of antiseptics (formalin, benzene, toluene, chloroform, etc.). In most cases, enzymes are most fully extracted at pH = 5-7.

In order to obtain concentrated extracts with small losses of enzymes with the biocrude, it is necessary to use special extraction plants. Until recently, diffusion batteries were widely used for this purpose. They can produce extracts with a dry matter content of 7 to 14 %, depending on the crop species, the medium and the amount of extract extraction.

The diffuser is a cylindrical-conical container equipped with a jacket. A dry or wet surface culture is placed in the central cylindrical part of the apparatus and secured with two nets at the bottom and top. Extractant is fed into the diffuser from the bottom, usually water at 22...28 °C, and the diffuser is filled completely up to the level of the drain pipe. The water supply is stopped for a certain period of time (depending on the type of enzyme it can be from 30 to 60 min), then it is switched on again and fresh water is used to force the extractant from the diffuser into the diffuser until it is completely filled, and again the whole system is stopped for a certain period of time. In this way everything is repeated until in the last, eighth diffuser the extractant remains for another τ_8, i.e. if τ = 30 min, then ^ = 44(0.54-8).

After that, the finished extract is taken. Continuity of the battery operation is ensured by the presence of 10 diffusers in the system. After passing the first cycle of extraction through 8 diffusers, in diffuser No. 1 has already passed 8-fold extraction, there are practically no enzymes in the culture, so this (No. 1) diffuser is switched off for unloading, water is supplied to diffuser No. 2, at the end of the process diffuser No. 9 is filled with fresh culture, and the extract is taken from diffuser No. 9 after 30 min of infusion. By this time No.

1 is unloaded and washed, No. 10 is loaded with fresh culture, No. 2 is switched off for unloading, water is supplied to No. 3, etc. In fact, the diffusion battery operates in a continuous mode: each diffuser is periodically in head, intermediate and tail states.

But these units for extraction of enzymes from surface culture have low productivity, require a lot of manual labour and there is significant loss of activity.

Thus, research and search for the most perfect design for the extraction of enzymes from surface cultures of microorganisms in a continuous mode with minimal manual labour are constantly being carried out. A diffusion apparatus of S. M. Grebenyuk's design was successfully tested at fermentation enterprises of our country and worked for some time with good performance. Grebenyuk. Its extractor consists of a horizontal housing with a screw and a vertical housing with a screw. Fresh culture is loaded through a doser; the extractant is fed through a leak and then passes through the vertical casing and into the horizontal casing up to the sieve belt, where the extract is taken from the mushroom surface culture, i.e. the extractant flow is in a state of countercurrent with the microorganism culture.

The rotary extractors used in enzyme production are Koppch Opp rotary extractors, which consist of a stationary casing, inside which there is a rotor divided into 16-20 compartments, rotating around a vertical axis. Each compartment has a sieve bottom on which the crushed mushroom culture is fed. The rotor slowly rotates and sequentially passes four sections, each of which is wetted with water or extractant from the previous compartments, the extract is sucked out by a vacuum pump and fed into the next compartment to moisten fresh culture, where it is again selected and transferred to the third compartment, etc. At the end of one rotation of the rotor, the bio-crotch is unloaded and the compartments are loaded again with fresh culture.

There is now a tendency to make greater use of press-diffusion, which consists in pressing the culture after infusion, then infusing it again at a lower concentration of enzymes in the resulting extractant, pressing it again, etc. It is likely that this principle, once the apparatus has been successfully developed, will find wide application in industry.

20.8. Concentration of enzyme solutions

Extracts of non-surface culture microorganisms and filtrates of deep culture are unstable during storage. In order to obtain ready forms of technical preparations (P2x and G2x), they need to be concentrated. The most common method of enzyme preparation technology for this purpose is vacuum evaporation. Vacuum evaporation is used as one of the steps in the preparation of dry technical or purified enzyme preparations. Enzymes are very sensitive to evaporation temperature, so the main condition for concentrating enzyme solutions is that the process should be carried out briefly at low boiling temperatures so that the evaporated liquid is not heated above a certain critical temperature for the enzyme, i.e. so that the enzyme is not inactivated. It should also be borne in mind that the purer the solution, the less it contains associated substances, the more sensitive the enzymes are to high temperatures. In concentrated extracts from surface cultures, enzyme inactivation is much less, because the extract contains a very large number of protective compounds that prevent enzyme inactivation. The temperature dependence of the extract enzyme stability after concentration by vacuum evaporation is similar to the behaviour of enzyme solutions with stabiliser at different boiling temperatures.

In the process of concentrating culture liquid filtrates, the solubility of many compounds changes and their precipitation occurs, "the total dry matter content in the concentrate decreases by 11-20 %, the pH of the concentrate changes. Mineral salts, some organic substances and their decomposition products precipitate, nitrogen loss is observed as a result of ammonia entrainment.

When concentrating the culture fluid of various micro-organisms, the mineral composition of the resulting concentrate changes significantly: for example, when concentrating the culture fluid of B.mesentericus. it is clear-cut.

The content of calcium, copper and magnesium decreases most sharply, the content of zinc and manganese decreases noticeably. Such a change in mineral composition of the culture liquid affects the stability of enzymes in the process of concentration. At thickening of culture liquid up to 10 % of " substance content the amount of calcium decreases only by 5 %, and copper - by 75 %. It is known, for example, that copper has an inhibitory effect on enzymes, while calcium has a stabilising effect. For this reason, at the first stages of concentration there is an increase in the activity of enzymes, especially proteinases. At deeper concentration with a sharp decrease in calcium content, the activity of enzymes also decreases.

Most enzymes are very sensitive to thermal treatment and need mild concentration regimes. In subsequent studies it was found that not only the boiling temperature of the concentrated solution has a great importance, but also the temperature of the heating steam (coolant). So, even at very low boiling temperature (25...30 "Q there is a noticeable inactivation of enzymes (up to 12%), if the temperature of the heating steam is equal to ^{120} "^"At the temperature of the coolant 90 ... 100 ^"boiling temperature 35 ... 40 ^{O}C loss of activity does not exceed 10%. It should also be noted that the higher the temperature of the coolant, the more dry substances of the concentrated liquid precipitate, especially at high boiling temperatures.

Depending on the species of the producer, the culture liquid has a different chemical composition and contains a different complex of enzymes, so the thermal regimes of vacuum evaporation are specified experimentally in each specific case.

The total loss of activity during vacuum evaporation depends to a large extent not only on the mode of concentration, but also on the design of the apparatus. Apparatuses for the stage of vacuum evaporation have been significantly improved in recent years, the duration of the process has been reduced by dozens of times, which has led to a noticeable reduction in the loss of enzyme activity, and also allowed to slightly tighten the temperature regimes for concentrating enzyme solutions. In addition to tubular vacuum evaporators with different tube arrangements (horizontal, vertical and inclined), with built-in and remote heating surface, with the use of forced circulation, new designs of film evaporators, ultracentrifugal vacuum evaporators and plate evaporators have been created. Of particular interest are rotary film evaporators, where the evaporated liquid in the form of film moves along the inner wall of the apparatus. Blades mounted on a rotating rotor continuously direct its movement from top to bottom. The time it takes for the liquid to pass through the apparatus is a few seconds. *Alfa Laval* currently manufactures vacuum centrifugal evaporators. They are very compact, the contact time of the enzyme solution with the heating surface is extremely shortened (no more than 1 s), losses do not exceed 10 %, the productivity of these units from 800 to. 4800 l/h.

A film-type evaporation unit with a capacity of 800 l/h for evaporated moisture was created. 800 l/h for evaporated moisture. The contact time of culture liquid with the coolant is not more than 1 s, the temperature of heating steam is 60...80 ^{O}C. In order to increase the productivity it is possible to mount the installation of three modules, each of which works either independently or in series, or the first two modules work in parallel and are connected to the third one in series. For enzyme production, a centrifugal film-type vacuum evaporation plant with a capacity of up to 200 l/h and a evaporation temperature of 30...40 ^{O}C is of interest. Rotary evaporators with evaporated moisture capacity from 50 to 200 l/(m$^{(2)}$-h) have good technological indicators. French company *APV* manufactures plate vacuum evaporators with a capacity of up to 20 thousand litres/hour.

Despite the availability of high-performance vacuum evaporators, it is not possible to completely eliminate the disadvantages of the vacuum evaporation method (loss of activity, precipitation, etc.), so it is increasingly replaced by ultrafiltration.

20.9. Membrane methods for purification and concentration of enzymes

Depending on the driving force of the process, membrane methods are classified into diffusion-dialysis (the driving force is the concentration difference on both sides of the membrane), electromembranous - electrodialysis (electric potential difference), baromembranous - reverse osmosis, ultrafiltration, microfiltration (pressure difference).All these processes are used for processing of enzyme solutions, their choice is determined by the purpose of processing: for purification from low-molecular impurities at low productivity - dialysis, for desalting in intensive conditions - electrodialysis, for deep purification from impurities by simultaneous concentration - ultrafiltration, etc. Let's consider in detail the essence and technical solutions of each group of processes, which makes it possible to further choose the optimal variant for any specific task.

Dialysis is the first studied and industrialised membrane process, as it does not require complex apparatus and special membranes. Dialysis consists in the fact that if two solutions with different concentrations of a component are separated by a membrane, a natural diffusion process starts, which reaches equilibrium when the concentrations of this component are equalised on both sides of the membrane. The intensity of the substance Q_c transfer through the membrane is determined by the diffusion coefficient of this substance in the membrane material. D_c and is proportional to the difference in concentration of LS. This dependence can be written in the form of a comparison: Q_c = D(c) - S - LS, where S is the membrane area.

Accordingly, the greater the difference in the diffusion coefficients of two components in solution, the better they are separated by the membrane. It is clear that protein molecules (high molecular weight substances), organic and inorganic low molecular weight molecules and ions of accompanying components (sugars, amino acids, mineral salts, etc.) are practically completely separated by the membrane due to huge differences in diffusion coefficients.

As dialysis membranes, films made of cellulose (cellophane, cuprophane) and other synthetic polymers are usually used. The process is carried out either by the flow-through scheme, when the initial enzyme solution is constantly pumped by one side of the membrane and the adia- lising liquid (usually water) - by the other, or by the semi-flow-through scheme, when the enzyme solution is placed for a certain time in bags from the dialysis membrane, which are constantly washed with water. In this way, the bulk of the associated low molecular weight impurities can be removed and the activity of the enzyme solutions can be increased several times over in conversion to dry matter.

The process of dialysis applied to the purification of enzyme solutions has a number of significant disadvantages. Firstly, dialysis may result in "loss" of enzyme as a result of leaching of metal ions included in the enzyme molecule or compounds stabilising the enzyme, or fragments of the enzyme itself, e.g. its prosthetic group. Secondly, during dialysis against ordinary tap water there can be a loss of enzyme activity as a result of metal ions - enzyme inhibitors - getting from water into the enzyme solution. It should also be noted that in the process of dialysis simultaneously with purification there is a strong dilution of the enzyme solution due to the penetration of water under the action of direct osmosis forces into the dialysed solution. The volume of the dialysed solution increases by about 20-25 %, and taking into account the active removal of ballast substances, dialysis results in very dilute enzyme solutions. Thus, this method of purification of enzyme solutions from ballast substances is almost never used in modern enzyme production. Sometimes it is used in laboratory research and in the preparation of highly purified enzyme preparations.

Electrodialysis. If in the process of enzyme purification the task is to remove electrolyte impurities, i.e. organic and mineral ions, electrodialysis is sometimes used. The essence of this membrane method is that the transfer of ions through the membrane is intensified by means of a constant electric field, and the membranes are made of special ion-exchange materials based on synthetic polymers.

The electric potential is supplied to the apparatus through two electrodes placed in corresponding electrode chambers. Both chambers are separated from the working desalting chamber, where the initial solution is fed, by ion exchange membranes, on the cathode side - anion exchange membrane, on the anode side - cathron exchange membrane. At work of the apparatus cations under the action of a constant electric field are displaced to the anode, meet on the way cation-exchange membrane, pass through it into the electrode chamber and in the form of a weak solution of alkali are removed from the apparatus.Correspondingly.behave and annons, leaving the apparatus in the form of a weak acid solution. The desalted enzyme solution (dialysed) is discharged from the working chamber.

The electrodialysis method is always continuous and more energy intensive than dialysis. It has the same disadvantages for the treatment of enzyme solutions. In addition, electrodialysis cannot be used for the isolation of enzymes having, for example, a quaternary structure, which is formed with the participation of metal ions, as well as for the isolation of metalloenzymes, which usually lose their activity during electrodialysis (α-amylase, ß-ra-lactosidase and others).

Baromembrane methods. These methods of separating liquid mixtures are often referred to the process of conventional filtration, but the resemblance is only external, due to the fact that the driving force is the difference in pressure- . In reality, semipermeable membranes are used to separate true solutions, i.e. homogeneous systems, whereas filtration can only separate suspensions, i.e. the solid phase can be separated from the liquid phase.

At the same time, considering membrane methods as filtration at the molecular level, it is possible to construct a conventional spectrum of filtration, placing membrane methods - reverse osmosiaiofiltration, ultrafiltration and microfiltration - in the form of some row and supplementing it with conventional mechanical filtration in order of increasing size and molecular weight of retained particles.

Today, baromembrane methods are widely used in food, pharmaceutical, and chemical industries. In particular, no modern enzyme production can do without ultrafiltration purification and concentration of the product.

Dissolution of a substance in a solvent is possible only when they have affinity for each other, i.e. when solvation of molecules or ions of the substance to be dissolved by solvent molecules occurs at the level of intermolecular interaction. In the case of aqueous solutions, the process is called hydration. Since the water molecule is a tiny dipole, its binding energy to the solute particle is greater the more charge the particle carries. It is clear that the greater the charge of the ion, the more water molecules will be bound to the ion in the form of a multilayered hydrated shell.

It is the formation of hydrate shells that explains the phenomenon called "direct osmosis". If a solution of any substance is separated by a semipermeable membrane from the volume of pure solvent, there will be a one-way transfer of solvent molecules (in this case water) into the solution, where they complete the hydrate shells. The higher the concentration of solute on the left, the more water molecules must pass through the membrane into solution. This transfer is expressed quantitatively by the osmotic pressure ($P_{o)}$:

$$R_o = CRT,$$

where C is the mass concentration of the dissolved substance; *R is the* gas constant; *T is the* absolute temperature

If the osmotic pressure ($P_{0)}$ is greater than the hydraulic pressure (P_d). then direct osmosis occurs, if $P_o = P_d$ then diffusion across the membrane stops.

If now a working pressure higher than the osmotic pressure, $P_r > P_0$, is applied to the solution, the transfer of water molecules from left to right will start, i.e. dehydration of the solution, concentration of the dissolved substance and obtaining pure water in the right half of the vessel. This mechanism is called reverse osmosis.

Reverse osmosis is similar in mechanism to Culgraphy Ultrafiltration is increasingly used in the technology of enzyme preparations from year to year. The data on the advantages of purification and concentration by ultrafiltration are very convincing (Table 12).

Table 12

Stages of enzyme purification of preparations

Separation method	Dry matter concentration, %		Costs for removal of 1 m^3 of water, US$.
	Initial	In concentrate	
Centrifugation	1-2	10-15	0,15-0,9
Gel filtration	3-5	diluted	6-30
Drum drying	30	100	7,5
atomising	10-20	100	15
lyophilic	10	100	60-90
Precipitation with ethyl alcohol or salts	1-2	Miscellaneous	1500
Ultrafiltration	1-10	10-50	0,15-0,30

Indeed, it follows from Table 12 that ultrafiltration has the ability not to inactivate enzymes and requires minimal energy inputs

The ultrafiltration rate will be the higher the difference between the working.hydraulic(*P"*) and osmotic pressure.

However, there are still differences between the processes of reverse osmosis and ultrafiltration. Thus, in reverse osmosis, the separation of low molecular weight substances occurred at working pressures up to 0 7-14 MPa, because the osmotic pressure P_o in these solutions is high. At reverse osmosis membranes with very small pores (from $1 - 10^{-4}$ to $2 - 10^{(-3)}$ microns) are used. In ultrafiltration, high and low molecular weight compounds are separated and the aim is to obtain a concentrate of high molecular weight compounds (e.g. enzymes). The operating pressure in this case is low (0.07 to 0.7 MPa) because P_o is small. The pore size of the membranes is much larger, from $3.10\text{-}^3$to 150.10-^3μm.

However, these distinctions are rather conditional. The mechanism of the processes of reverse osmosis and ultrafiltration is still not completely clear.

The mathematical description of the membrane separation process is based on the model of viscous flow through pores (Poiseuille equation) and the model of diffusive mass transfer (Fick's law). It is assumed that if the membrane pore size is less than 3 - 10 ^{3} µm (reverse osmosis), the process obeys Fick's law; if the pore size is greater than 3 - 10 ^{3} µm (ultrafiltration), the process obeys the Poiseuille equation.
This is the principle of any baromembrane process. The only differences between them are the pore size of the membrane used and the applied pressure.
Since the osmotic pressure of protein solutions is small, 0.3-0.6 MPa is sufficient for the process, and the pore size of the membrane should be 10-50 nm.

20.10. Use of microbial enzymes

The use of microbial enzymes in various fields of national economy is very promising. At present, enzyme preparations derived from microorganisms are used in various fields of industry, agriculture and medicine.
In brewing and winemaking, mushroom amylase preparations are used to replace malt. This makes production cheaper and reduces grain consumption. Amylases are also used to produce soluble starch, dextrin and molasses. Fruit and vegetable products produced with amylases contain more sugar and are better digested, especially by children. Amylases accelerate the ripening of dough and improve the quality of bread.
In the confectionery industry, yeast invertase (sucrose) is used to convert sucrose into glucose and fructose. It prevents sucrose crystallisation at high concentrations.
A complex of enzymes (cytases) of fungi, which break down substances of plant cell walls, is used to improve the extraction of their contents (juice, essential oils, fats, starch).
Mushroom pectinases are used to clarify fruit and berry juices, increase grape juice yield in vniodelia, and in coffee production. The use of pectinases is especially effective in the production of juice from fruits and berries containing a lot of pectin (blackcurrant, gooseberry, plum).
Mushroom glucoamylase is used in the brewing industry to remove dextrin residues from beer. Glucose isomerase is used to produce glucose-fructose syrups to replace sucrose, which is important for improving diets as the use of large quantities of sucrose is harmful to humans.
Lactase is used to produce lactose-free milk. After this treatment, it has a higher level of flavour. In addition, some people cannot consume milk because it contains lactose, which causes allergic reactions. Lactase is also

used to produce sugars (glucose, galactose) from whey containing large amounts of lactose.
Glucose oxidase of mushrooms plays an important role, as it allows food products to free themselves from glucose residues and molecular oxygen and thus increases their shelf life. Glucose oxidase is added to egg powder, mayonnaise, and beer for long-term storage. With the help of this enzyme, the oxidation of ascorbic acid is slowed down during the processing of fruits and vegetables. The use of enzymes facilitates the production of gluconic acid.
Asp. niger catalase is used in the food industry to remove residues of hydrogen peroxide, a sterilising agent in the production of food concentrates, sterile milk, and melange.
Cellulase preparations are used for saccharification of potato pulp, extraction of starch from potatoes and grains, increasing the yield of agar-agar from algae, preparation of vegetable paetas, removal of citrus peel, and production of reducing sugars from plant materials. This method of producing sugars can be cheaper than using starch as the starting substrate.
Proteolytic enzymes of microbial origin replace renin in cheese making to produce clots. They are used to tenderise meat, accelerate the ripening of fish during salting, in winemaking and brewing.
Lipases are used in the production of whole milk powder and in cheese-making to accelerate the ripening of cheeses and give them a specific flavour and aroma.
Application of enzymes in the textile industry. In the textile industry, pectolytic enzymes of microorganisms have long been actively used for processing flax straw and obtaining fibre from it. Today, thermal linting of flax is used in flax mills. The main microorganisms involved in the process of linting are recognised as anaerobes of the genus, which occur during linting, lead to the destruction of pectin substances of flax straw and the release of flax fibre.
To accelerate the process (usually 2.5-3 days) and improve the quality of fibre, the enzyme preparation pectoclostri- din GZh - filtered and dried culture liquid of *Clostridium* sp. The preparation allows to accelerate the lobe process 2-2.5 times and to obtain fibre with high mechanical strength. Amylolytic preparations are used to remove adhesives from tissues (desliming). Some proteinases, in particular protosubtilin, are used to de-glue silk (removal of sericin) and release silk fibres consisting of fibroin. Lipase preparations are used to free silk fibres from fat.
Other industrial applications of enzymes. In the leather industry,

microbial proteinases are used for dehairing hides and softening leather. The use of a complex enzyme preparation consisting of protease and lipase speeds up the process and produces high quality wool.

The use of microbial enzymes in the production of detergents is becoming increasingly widespread. Usually enzymes *of Vas. subtilis* enzymes possessing proteolytic, amylolytic and lipolytic activity; the preparations are used in combination with surfactants. Detergents containing enzymes reduce the duration of washing and increase the preservation of fabrics, as the treatment is carried out at 40...60 ^{O}C (not higher).

Application of enzymes in agriculture. In this area, enzymes are used in animal diets and in the processing of feed to increase its digestibility

The surface method of cultivation of *Asp. oryzae* in our country produces the preparation amilorizin, a dried culture of the fungus containing "-amnlase, dextrnnazumaltase, glucoamylase and pro-theinase. Glucavamorin preparation - dried culture of *Asp. awamory* grown on bran, contains "-amylase, dextrnnase, maltase, glucoamylase, acidic proteinase and hemicellulase. The preparation amylo-subtilin contains "-amylase-protease, ß-gluconazun lytic enzymes, which are also included in protosublitin GZh and xylavamori na GZh, containing also hemicellulase and pectinase.

Medical applications of enzymes. Microbial enzymes are used in various fields of medicine as therapeutic agents, as well as in clinical analyses. In the treatment of inflammatory processes and burns, proteinase preparations are used to destroy necrotised tissues and cells, promoting rapid healing.

Bacterial *L-asparaginase* is used in the therapy of malignant neoplasms, which converts *L-asparagine* required by leukaemia cells into *L-aspartic* acid, as a result of which tumour growth is significantly slowed down. The proteinases terilitin and streptokinase, which are of microbial origin, have thrombolytic properties

In replacement therapy, i.e. in case of impaired synthesis of some enzymes in the human body, separate enzyme- ™ and complex enzyme preparations are used. For example, when the function of the pancreas is impaired, a complex preparation containing proteinase, amylase and lipase is used. When the ability to synthesise lactase and glucoamylase is lost, these enzymes are also used, obtained from microorganisms.

In some cases of digestive disorders, enzyme complexes (α-amylase, cellulase, lipase and protease) are used.

The use of microbial enzymes in medicine is very promising and will undoubtedly expand, despite the fact that enzyme preparations for medical purposes require additional costs for their purification from pyrogenic substances, toxins, as well as for testing for antigenic effects and allergic reaction of the organism.

Use of enzymes in chemical analyses. The use of enzymes as chemical reagents constitutes a special field of analytical and preparative biochemistry. Examples include the determination of cocarboxylase in blood by *Asp. niger* proteinase, *which* breaks it down to release thiamine. The amount of free thiamine is used to calculate the amount of enzyme.

Galactose is determined using galactose oxidase from the fungus *Potysporus ciracmatus*. The enzyme oxidises galactose, converting it to the corresponding lactone with simultaneous formation of hydrogen peroxide. When peroxidase and *ortho-dianisidine* are added, the colour of the colorimetrized solution turns blue.

The determination of orotic acid can be carried out by the corresponding *orotic* acid dehydrogenase from *Clostridium oroticum*. The enzyme catalyses the reduction of orotic acid by $NADH_2$ to form dihydroorotate and NAD.

Bacterial hyaluronidase cleaves gnalluric acid quantitatively to form an unsaturated disaccharide, forming N-acetylglucosamine groups, which give a red colouring when heated in alkali with para-dimethylaminobenzaldehyde (important for colourimetrics).

Bacterial ".fungal proteinases are used to cleave proteins, and the cleavage may be fragmentary. Subtnlysin, for example, cleaves only one peptide bond of this protein when exposed briefly to narnboyuclease. Yeast carboxydopeptidase progressively cleaves proteins, separating one by one the amino acids at the carboxylic or amine end.

The determination of B-amino acids (L-lysine, *L-arginine,* L-arginni *L-oryntine* B-ty- rosine, etc.) can be carried out using specific amino acid decarboxylases. The enzymes are synthesised by some bacteria (e.g. *Cl. welchii, Cl. séptica*) and are effective at low pH values. The carbon dioxide formed during the reaction can be measured and quantified.

Glutamine and asparagine are determined by deamination by specific *Pseudomonas* zr. Ammonia released is determined by colour reaction with phenolate in the presence of nitroprusside.

Such enzymes of microorganisms as chol- sterin oxidase for determination of cholesterol in blood, alcohol oxn dase for detection of alcohol, etc. are very promising. Enzymes are also used to study the primary structure of high-molecular compounds: nucleic acids, proteins, polysaccharides.

Use of enzymes in organic synthesis. A promising area of application of microbial enzymes is their use in the synthesis of various practically important compounds.

It was noted above that with the help of penicillinamidase, formed by some bacteria, it is possible to obtain 6-amnnopenicillanic acid - a substrate for the synthesis of semi-synthetic penicillins.

L-Aspartic acid is produced by *E. coli* aspartase. Both enzyme preparations and cells with increased permeability in immobilised state are used.

Some other practically important *L-amino acids*, such as tyrosine, 3,4-dioxyphenylalanni, can be synthesised in a similar way.

The use of amino acid acylases makes it possible to separate their L- and *D-isomers*. In Japan, the production of *L-apple* acid from fumaric acid has been established using bacterial fumarase. This acid is used as a substitute for citric acid in the food and pharmaceutical industries.

A number of transformation reactions leading to the production of steroidal drugs are based on the enzymatic activity of non-growing cell-^microorganisms.

Microbial enzymes can also be used to carry out multistage processes. For example, immobilised yeast cells are able to synthesise glutathione from glucose and phosphate, which requires the sequential action of several enzymes.

As.already.noted, a special field of application of microbial enzymes is genetic engineering. Even these few examples allow us to realise that the possibilities of using microbial enzymes for the production of new compounds and for other purposes are very diverse. It is obvious that in the near future the volume and number of microbial enzymes produced by the industry will be significantly expanded.

The production of enzyme preparations occupies one of the leading positions in modern biotechnology and is one of the industries whose production volume is constantly growing and whose scope of application is steadily expanding. Such rapid development is due to the fact that enzymes are highly active non-toxic biocatalysts of protein origin, which are widespread in nature, and many biochemical processes cannot be carried out without them.

Enzymes served as a basis for the formation and development of enzyme preparation technology as a science, as well as for the creation of industrial production of the most widely used enzyme preparations. Their application helped to significantly change, intensify and improve many existing technologies and to create fundamentally new highly efficient ones. The use of enzyme preparations of various degrees of purification allowed not only to improve the performance and productivity of various technological processes, but also to improve fodder production, to make more targeted and effective the action of synthetic detergents, to improve the quality of cosmetics, to create a whole arsenal of specific, sensitive and accurate analytical methods, to establish the production of drugs and prophylactics for the medical industry, etc.

An important and undeniable advantage of enzymes over chemical catalysts is that they operate at normal pressure, at temperatures of 20...70 °C, pH in the range from 4 to 9, and in most cases have extremely high substrate specificity, which allows them to target only certain compounds in a complex mixture of biopolymers. All this indicates that the production of enzyme preparations is one of the promising areas in biotechnology, which will continue to be intensively developed and expanded.

Chapter 21

BIOTECHNOLOGICAL PRODUCTION OF FOOD ADDITIVES

21.1. General characteristics of food additives

Food additives are substances added to foodstuffs to give them desirable properties such as aroma (flavourings), colour (colouring agents), shelf life (preservatives), taste, consistency.

The overwhelming majority of products consumed by city dwellers contain food additives, which are labelled with the code with the letter "E". Almost any product offered to us in supermarkets - breadcrumbs, ice-cream, juice, semi-finished products, an ordinary loaf of bread - almost always has the letter "E" on the package. Whether it is possible to eat such food is up to the buyer, who, of course, is not informed about the possible harmfulness of a certain food additive contained in this product. Thus, before buying anything in the supermarket, you should have special information.

Food additives are an essential part of almost all convenience foods and many other perishable and long-life products. Food additives can improve the flavour of a product and extend its shelf life. But apart from all the positive qualities of food additives, they are mostly harmful to our health. On average, during one year a person consumes about 5kg of various food additives with food. In many countries, some of them are banned altogether.

The following substances are prohibited in Belarus: E103, E106, Elli, E121, E123, E126, E130, E181, E216, E217, E240, E924, E924a.

There is a perception that food additives are only harmful to the human body. But it should not be forgotten that many of them keep food fresh and healthy for consumption. Sometimes food additives are simply indispensable. According to some calculations, they perform about fifty different functions. Eleven groups of the most well-known additives can be distinguished. These include: nutritional additives, or natural food components; additives that preserve the freshness of food; additives that facilitate the processing or manufacture of food, as well as preservatives: seasonings, colouring agents, thickeners or texturants, sweeteners, fillers; additives that can reduce the caloric content of food, etc. The use of food additives can be used to reduce the caloric content of food. As a result of the use of food additives in some countries it has been possible to completely eliminate diseases caused by the lack of any substance in food. For example, goitre develops due to lack of iodine in the body, scurvy - due to lack of vitamin C, rickets - due to lack of vitamin D, calcium, phosphorus. In developed countries, such food elements as fats, carbohydrates, proteins and fibre are added to food products to improve their nutritional value. This is done to improve the nutritional value of the food consumed.

E100-E182- colouring agents;

E200-E200 are preservatives;

E300-E399 - antioxidants (antioxidants);

E400-E449 - consistency stabilisers;

E450-E499 are emulsifiers;

E500-E599 - acidity regulators, leavening agents;

Eb00-Eb99-$_y$flavour and aroma enhancers;

E700-E^{800} - spare indices for other possible information;

E900 et seq. are antiflaming agents, bread improvers, etc.

Additives are divided into natural and synthetic. Natural additives are produced only from natural raw materials such as herbs, spices, fruits, vegetables, meat, yeast, tree bark, fungi and even insect pests. However, the distinction between natural and synthetic additives is rather arbitrary, as it concerns not so much their composition as the way they are produced. Natural additives are not necessarily safer than artificial ones. They often contain more chemical impurities.

The method of obtaining food additives from plant raw materials containing vitamins, mineral salts and plant fibres can find wide application in food and medical industry. It consists in the following: the prepared plant raw material is pre-crushed, dried in certain conditions to a humidity of no more than 10 %, after which the resulting mass is cooled with a refrigerant until the temperature in the whole mass reaches -80...-165 ^{O}C with its subsequent grinding to particle sizes of 50-600 microns, then the finished product is packed. It can be tabletised before packaging. This allows to obtain a biologically active additive from traditionally used food raw materials, having high biological activity due to the most complete preservation of biologically active substances and their increased absorption by the body. Commodity examination of food additives is carried out at the stage of manufacture and at all stages of their movement, one of which is the creation and analysis of the technology of selection and introduction of food additives or their complex into the product, taking into account the features, chemical composition, functional properties of food additives and raw materials, the nature of action, type of product, technology, in some cases - packaging and storage.

21.2. Natural food colours

Relatively cheap synthetic colouring agents have found very wide application in the food industry. However, there are almost no absolutely harmless ones among them. Moreover, many imported colours are potentially dangerous as carcinogens and allergens. In addition, they do not always successfully imitate the colour of the natural product. That is why recently more and more attention has been paid to natural colouring substances. Natural (natural) food colours are colouring substances extracted by physical methods from plant and animal sources. Sometimes they are chemically modified to improve their technological and consumer properties. A number of colourants are obtained not only by their extraction from natural raw materials, but also synthetically. Raw materials for natural food colouring can be berries, flowers, leaves, root crops, etc., including in the form of waste processing of vegetable raw materials at canneries and wineries. The technology of production of natural dyes is constantly improving, which allows to increase their durability and optimise their price. Natural dyes are usually extracted from natural sources, in the form of mixtures of different compounds, the composition of which depends on the raw materials and the technology of production, so ensuring the consistency of their quality is a very difficult task. The shades and durability of the same natural pigments in dyes of different manufacturers may vary. Properties of natural pigments can be corrected and improved with the help of natural antioxidants and technology of suspension, emulsification and microencapsulation, the use of which allows to significantly expand the scope of use of natural colours. Their production is one of the most complex areas of the food industry, in which only a manufacturer with in-depth science and a highly developed production base can achieve a high level of quality.

The dye content of raw plant material depends on climatic conditions of growth and time of collection, but in any case it is relatively small (usually a few per cent or fractions of a per cent). The amount of other chemical compounds - sugary, pectin, protein, substances, organic acids, mineral salts, etc. - may exceed the colouring content by several times. These substances are not dangerous to health, and often even useful for it, but by their presence they reduce the intensity of colouring of the finished product. In the production of natural food colouring preparations, the by-products are eliminated.

Modern technology makes it possible to produce natural food colour preparations with defined properties and a standardised content of the main colouring agent. Consumers intuitively associate the appearance and quality of a product as a whole. This is why food colours have become the most important ingredients on which the "product appearance" depends.

Natural, colouring agents can be water- and fat-soluble, which broadens the scope of their application. Moreover, many natural colouring agents are dietary aids. There are a large number of natural dyes such as for example red sandalwood, quercitron, carmine, sepia, kampeshe wood, etc. Campeche tree grows in South America, sandalwood grows in South Asia, sepia is extracted from cuttlefish, carmine from cochineal (tiny insects).

Plants are mainly used in dried form. Drying is not done in the sun or in the oven, but in the shade in the open air, on stretched ropes or cloth. The colouring substances contained in plants are very rarely found in pure form, and the following methods and techniques are used to obtain them: first of all, the raw materials are pulverised by cutting, crumbling or grinding; the degree of pulverisation depends on the properties and qualities of the material itself, but in any case the pulverisation must be uniform. The material is then subjected to leaching, which consists^ of boiling in rain, river or distilled water, with a little potash or soda added to the water for a more complete extraction of colours. As the water boils off, add fresh water and so proceed until the material is completely leached, ie until the water will no longer be coloured. The decoction is then boiled for evaporation, i.e. to obtain a thicker colour. Let it stand and filter it through a sieve with a piece of cloth on it.

If you boil dried onion peels in water, you get a *brown dye* of various shades, from almost yellow to dark brown. Another source is the dried rind of joster.

Caramel colouring agent (E150 a-d) is produced by thermal processing of various types of sugar substances. It is brown in colour, well soluble in aqueous phase, not sensitive to light and temperature.

In recent years, the development of the market for ice cream, yoghurt, cottage cheese and fermented milk products has led to the emergence of many new original products. It is also possible to vary the shades of the final product by changing the dosage of the colouring agent or by using two or more colouring agents at the same time. For example, to obtain lime colour, it is recommended to use copper chlorophyllnium dye and curcumin at the same time.

Fat-soluble natural colouring agents are primarily used. by producers of chocolate glazes, chocolate. They have a fat component in their composition, so their introduction into fat-based products is much more technologically advanced than synthetic colouring powders (no high-speed mixers are required, etc.). The use of natural fat-soluble colouring agents enables the manufacturer, using natural raw materials, to obtain a wide range of shades for the products.

So, modern production is unthinkable without colouring agents. But according to market trends, it is natural ingredients that are needed nowadays. Market research confirms this. It is true that the production volumes of synthetic dyes are still Z.5 times higher than those of natural dyes. However, the growth rate of natural dyes is higher than that of synthetic ones.

Red dye can be obtained from the stem of St. John's wort (the decoction should be acidified), the root of St. John's wort, decoction of alder, bark, placed in water for a few days. Beautiful dye can be extracted from the roots of horse sorrel, but in this case do not forget to add to the finished decoction of a little aluminium alum, otherwise the colour will be dull. Red shades are represented by various anthocyanins E163, carmines E120, as well as beetroot juice. Shades of anthocyanins can range from red to red-violet. They can be obtained from grape skins, black carrots, elderberries, etc. The dyes are water-soluble, light- and heat-stable, the shade depends on pH. Red dyes of flavonoid type (in particular anthocyanins) are obtained by elution with 50% acetic acid or extraction with hydrochloric acid and ethyl acetate. Carmine is obtained by extraction from the dried and shredded female insects of *Coccus Saette, an* insect species found on cacti in South America and Africa. The carmine pigment has good resistance to light and high temperature. Carmines can be represented by both water-soluble and fat-dispersed forms with shades of red to reddish-purple.

The source of beetroot juice colouring agent (E162) is *Beta vulgaris* beetroot. The main pigments contained in the juice are betanin and valgaxanthin. Beetroot juice gives products a colour from pink to beautiful blue. The main disadvantage of beetroot juice is its poor heat stability. However, this is not a key point for the dairy industry. The colouring agent is available either as a water-soluble liquid or as a powder.

Blue dye is extracted from the roots of elecampane. For this purpose, the roots must first be kept in ammonia ammonia solution. Blue dye can also be extracted from the flowers of larkspur and bird's-herb root

Green colouring agent (chlorophyll E140) is a natural colouring agent made from the green mass of plants (nettle, spinach) and gives olive shades to products. Copper complexes of chlorophylls, chlorophyllins (E141) are used to give products bright green shades. A green dye is extracted from the leaves of shamrock (a medicinal herb). A less bright, but nevertheless beautiful grey-green dye can be obtained from the leaves and stems of cuffs; before preparing a decoction, they must be thoroughly crushed.

G~ *colouring* agent is produced by many plants: dyer's wort, hazel (bark), alderberry (bark, leaves, berries), bramble (flowers). Barberry fruit produces a yellow dye with a lemon-tint.

Blueberries and jellyberries, as you can easily guess, contain a *phyllo-colouring agent.* It is not very stable, but it can be useful for watercolours.

The orange *colouring* agent can be extracted from celandine and celandine leaves. Carotenoids are obtained by adding ether-acetic tonic and acetic acid to vegetables, followed by diethyl ether.

Black dye can be extracted from decoction of berries and roots of funnelcorn. But there is another, simpler way: add iron sulfate to one of the previously obtained decoctions. Almost all of these decoctions contain tannins like tannin, and in the presence of divalent iron salts they become black.

The natural off-colour of cooked meat is unfortunately not very appealing to most consumers, so the meat industry makes extensive use of colour-correcting ingredients. The meat industry consumes 13 % of the natural colouring agents produced worldwide. The ideal food colouring agent for this industry should be stable, harmless at recommended dosages and conditions of use, not adversely affect the properties of the food product, not react with the main and associated components of the product being coloured, have a high colouring power, be easy to use, and be reasonably priced.

The natural yellow-orange colouring agent annatto (E160b), which belongs to the carotenoid group, is widely used in the meat industry. Annatto is obtained from the seeds of the orleans tree *Βιχα oreUana.* It is native to tropical America. This tree has long been cultivated in tropical countries to produce a natural orange coloured dye. Annatto has a mixture of carotenoids, the main one being bixin. Bixin is fat-soluble and its derivative norbixin is a water-soluble pigment. Annatto has good heat and light fastness. Special acid-resistant forms of the colouring agent have been developed for use in products with acidic pH. The water-soluble form of annatto (Norbic syn pigment) has the ability to form strong complexes with proteins. Due to this ability, annatto is well suited for colouring the casings and surfaces of smoked products. To obtain a red or bright red-brown shade it is necessary to add 50-100 g of kraev-tel per Yul of water (the dosage is specified for a colouring agent with pigment concentration of about 1%). The shell should be immersed in the prepared solution for 2-3 hours or more, depending on the desired colour intensity. When dyeing meat products, dip them in the dye solution dissolved in cold soft water for 2-5 minutes until the desired colour is obtained. Для окрашивания копченостей этот краситель иногда комбинируют с натуральным коричневым красителем карамель, присемизчетырехее видов в мясной промышленности используют только E150C в E150Й.Наиболееподходящим для сочетания с аииатто и по оттенку, и с технологической точки зрения является E150c. If is necessary to give smoked meat a slightly brighter orange colour, a little carmine is added to aiatto.

To obtain a golden yellow colour, annatto is mixed with the natural colouring agent *curcumin* (Turmeric) (E100). The natural colouring agent curcumin allows foodstuffs to be coloured yellow. *Curcumin* (Turmeric) (E100) is obtained from the rhizome of the spicy turmeric plant *Curcuma longa L. of* the ginger family, native to India. Curcumin is fat-soluble in nature and disperses in water when emulsifiers are added. Curcumin is a strong antioxidant, improves digestion, helps normalise intestinal microflora, binds free radicals, helps fight age-related diseases, prevents cancer, is an antibacterial agent, promotes liver detoxification, prevents the formation of cataracts and gallstones, stimulates the gallbladder.

One of the most widely used natural colourants in the food industry is *paprika* extract (E160c). Paprika is a colouring agent obtained by extraction from the red sweet pepper *Capsicum appium, which* grows in Europe and North America. Paprika contains the carotenoid pigments capsanthin, ß-carotenne and capsorubin. Depending on the pigment content, the dye can range in colour from red to orange. It improves digestion and has antiseptic properties. Pigmentspaprnki fat-soluble in nature, but on their basis are also obtained dispersible in water dyes. Depending on the production condition and composition of water-dispersed paprika, this colouring agent may behave differently when mixed with water. The cheapest forms of water dispersible paprika are dispersed in water with careful mixing; colouring agents that can be easily mixed with water have a slightly higher price. This colouring agent is most in demand in the meat industry and is widely used in the production of spice mixtures and seasonings. Paprika is used in almost all areas of the food industry. A well-purified colouring agent, if the recommended dosage is observed, does not impart any perceptible extraneous flavours and odours to products. Paprika oleoresins are not visibly sensitive to the pH of the medium, but without special protection, paprika pigments are not very resistant to light and high-temperature treatment. Natural antioxidants (tocopherols, rosemary extract, etc.) are used to increase the light and heat resistance of paprika. In meat semi-finished products, water- and fat-soluble forms of colouring agent (containing 1.5% of colouring pigment) are usually used in quantities of 0.05 to 0.1%.

The most commonly used natural colouring agent for giving meat products a natural red colour is *carmine* (E120), a natural colouring agent of animal origin, which is a complex compound of carminic acid with calcium and aluminium. Water- and fat-soluble colourants can be produced on the basis of this pigment, and it is also used as a water- and fat-insoluble complex compound with calcium and aluminium. Carmine varnish has a lighter shade of red and is easily mixed with water and fat-based products. This form of this colouring agent has a low migration capacity. Carmine has proved to be one of the most stable food colouring agents: while giving sausage and delicatessen products a natural, juicy colour, it is not visibly sensitive to light, oxidation and temperature treatment. Carmine is used both in the production of raw smoked and cured products and in the manufacture of sausage products subjected to heat treatment. Carmine in water-soluble form is the only natural colouring agent used for injecting cooked ham and for colouring casings in different shades of red (often in combination with annatto). This dye makes it possible to obtain a stable colour in the production of meat products, to regulate the degree of their colouring according to the type and wishes of the consumer, and to improve their appearance

"L (E161) is obtained by extracting the petals of *Aztec Mangolds*, which grows in South America. The colour is 318

The colouring agent is a water-dispersed extract that is soluble in fats and oil. It gives the product various shades of yellow colour from lemon to golden. Lutein is less sensitive to oxidation than other carotenoid colourants, stable to heat, light stable.

21.3. Preservatives, antioxidants, food stabilisers, emulsifiers, flavour enhancers, texturants, sweeteners, fillers

Preservatives. The most used preservatives are: *table salt, ethyl alcohol, acetic, sulphuric, sorbic, benzoic acids and some of their salts*. Preservatives can be divided into preservatives proper and substances with preserving action. The former affect directly the cells of microorganisms, the latter have a negative effect on microbes, mainly by reducing the pH of the medium, water activity or oxygen concentration. Each preservative has its own spectrum of action.

The accumulation of preservative in the product can occur not only by its introduction from outside, but also due to chemical changes occurring in the raw material as a result of the activity of microorganisms. Fermentation of cabbage, pickling of cucumbers and other vegetables is based on lactic acid fermentation of sugar, which results in the accumulation of lactic acid, which is a natural preservative.

As the population grows, preservatives become more and more important, as it is much easier to achieve a 10 per cent increase in food stocks with preservatives than by expanding agricultural production.

Nisin (E234) is a naturally occurring antibiotic, a product of lactic acid bacteria. It actively inhibits the growth of heat-resistant Gram-positive spore-forming bacteria, prevents the formation of botulinum toxins. It is used in the production of cheese and other dairy products, canned vegetables, meat and fish, as well as in winemaking, brewing and baking.

MoLOchnaya ***acid*** (E270) is formed in the process of lactic acid fermentation and is widely used as a preservative in the manufacture of cheese and a number of DRU™h dairy productsMuravaic acid and its derivatives (E236-238) are used mainly in the preservation of vegetables and the production of soft drinks.

Antioxidants. Additives that preserve freshness include primarily antioxidants. They are used in oils and packaging materials to prevent rancidity. Chelating agents and ejectants are also used. These prevent interaction between metals and food components, minimising discolouration and loss of flavour and aroma. A number of substances are used to prevent fruit from darkening on the cut surface

Like preserving agents, antioxidants are used for extending the shelf life of food, mainly fat-intensive products. Their action is based on the inhibition of oxidation reactions of food components. Oxidation occurs under the influence of air oxygen, light, temperature, technological factors of production. First of all, lipids and their compounds, vitamins and other biologically important nutrients are oxidised, which reduces the nutritional value of the product. The final oxidation products have a negative effect on organoleptic properties and can be toxic to the human body. For example, oxidation of lipid components leads to the formation of hydroperoxides, which, also oxidised, form such toxic compounds as aldehydes, ketones, certain fatty acids and numerous products of their polymerisation.

Natural *antioxidants* include tocopherols (vitamin E), ascorbic acid (vitamin C), flavones (quercetin), gallic acid esters, etc.

Food stabilisers. This is a special group of additives used in different branches of food industry, the main purpose of which is to form and preserve the consistency, texture, shape and consumer qualities of products of dairy, meat processing, bakery and confectionery production.

There are three main groups of food stabilisers: pectins, carrageenans and gum. They are all derived from natural substances, although recently the volume of global food production has necessitated the industrial synthesis of some types of food stabilisers. Food stabilisers are not hazardous to health

and are a very important aid in increasing global food production. The raw materials for them are apples, citrus fruits, wheat, corn, seaweed, etc. They are used in the production of food stabilisers.
resins of various terrestrial plants, etc. Some types of stabilisers are products of microbiological processing.
(Pectin)^ (E440) is a natural jelly-forming substance found in fruits and many types of vegetables Pectin is usually obtained by extraction from citrus fruits or apples. Its peculiarity as a gel-forming agent is its ability to form gels in aqueous solutions only in the presence of a certain amount of sugar and acid or calcium ions. Pectins are used in confectionery, dairy processing and meat processing industries
The other group includes three types of gum: *guar* gum (E412), *xan- tan gum* (E415) and *carob gum* (E410). Gums serve as thickeners, stabilisers, gel-formers, capsule- lyring agents. They are widely used in the production of processed cheese, ice cream and dairy products, fruit and vegetable preserves, raw smoked sausages, sauces, ketchups, mayonnaise, bakery products, canned fish, low-fat margarines and spreads. Gums are also used in conjunction with other thickeners and gelling agents to regulate the process. For example, guar gum is used in combination with carrageenan to produce cheese.
Carrageenan (E407) is a natural thickening agent obtained by processing red seaweeds *of the Rhodophyceae* class, which grow almost all over the world's oceans, on underwater rocks at depths of up to three metres. Carrageenans are also widely used in the above-mentioned food industries.
Emulsifiers. Substances capable of forming and stabilising an emulsion, enabling the creation and preservation of a dissociation of two or more miscible substances. Gums, saponins, lecithin and other natural substances were first used as emulsifiers. Nowadays the list of emulsifiers has expanded mainly due to synthesised preparations. Emulsifying ability of this group of substances is associated with their surfactant properties, so the term "emulsifier" can be considered as a synonym of the terms "emulsifying agent" and "surfactant". The main field of application of emulsifiers and stabilisers is the oil and fat industry. Thus, for preparation of fats used in bakery and confectionery production, emulsifiers are allowed: T-1 - mono- and diglycerides of fatty acids; T-2 - a product of esterification of polyglycerol with saturated whey acids (C16 and C18). They are added in the amount not more than 2 g/kg per 1 kg of product. For these compounds the permissible daily dose (PDD) is 125 mg per 1 kg of human body weight. In addition to their main function, emulsifiers are used to evenly distribute fat-

soluble substances and compounds in water: flavouring agents, essential oils, spice extracts, etc.

Flavour enhancers. Fresh vegetables, meat, fish and other products have a bright flavour and aroma due to the content of nucleotides in them. During storage and industrial processing the amount of nucleotides decreases, which is accompanied by loss of taste and aroma of pro-

Monosodium glutamate (E621, or sodium salt of glutamic amino acid) is a food additive intended to enhance taste sensations by increasing the sensitivity of tongue receptors. Monosodium glutamate contains no nutrients. When we eat food, it "tricks" our brain into thinking that it tastes good.

Texturants. These are various additives designed to improve the texture of food products. Calcium compounds make tinned tomatoes denser and firmer. Phosphates improve the flavour of tinned pears, making them more tender. Pyrophosphates improve the texture of instant puddings and dairy products. Emulsifiers give stability to water and oil emulsions in salad dressings. Substances such as starch are used in a variety of applications to give products greater density. Leavening agents provide the appropriate texture for baked bakery and confectionery products.

Sweeteners. Natural sweeteners such as sugar have been known to humans for thousands of years. They have always been produced in large quantities. However, concerns about reducing caloric intake have forced a turn to non-food sweeteners. Five such substances are currently authorised for use in the USA: saccharin, aspartame, acesulfame, thaumatia iglicirisii A number of substances authorised in other countries are also under consideration. Aspartame and acesulfame are about 200 times sweeter than sucrose. Work is underway to develop new, more effective artificial sweeteners.

Fillers. The trend towards non-food sweeteners has led to the search for substances that can fulfil the role of sugars in drinks, jams, jellies and cured meats. Gelatinised starch has been used for centuries, and now a number of starch and cellulose derivatives have been produced. Polydextrose, also one of the sugar derivatives, is used.

Chapter 22

PRODUCTION OF LIGNIN

The presence of lignin in plants is usually associated with the presence of supporting "conducting tissues" (e.g. xylem). Lignin is an important component of the cell walls of these tissues. It is found in plants such as plaunas, ferns, holoderms and coleopterans, but it is not found in non-vascular plants such as fungi and algae. Mosses are an exception in that, although they do not have xylem-like cells, they nevertheless contain ligion-like compounds; however, there is some doubt as to whether they are true ligions.

"Lignin is contained in the cell walls of not only conductive tissues but also the heartwood, roots, fruits, buds, bark and cork. The lignification of the cell wall occurs after the deposition of polysaccharide components of the wall and towards the end of the cell growth period. The distribution of lignin in the wall is uneven; usually, the layers that were deposited first are the richest in lignin. Consequently, the medial lamina and primary wall undergo the greatest lignification, while the authoritative wall undergoes the least.

Lignification has a twofold significance. It strengthens the cell wall by forming a branched network throughout the matrix, thus anchoring the cellulose microfibrils more firmly. In addition, lignin protects the microfibrils of the wall from chemical, physical and biological effects.

An important property of lignin is its resistance to microorganisms. Only a few of them degrade lignin, and then only relatively slowly.

Lignin is a group of closely related high molecular weight polymers whose main, if not the only, building block is a phenylcropane residue. There are few known structural variants of the phenylcropane residue .There are many pathways by which residues are joined together. There is no order in the structure of the polymer. The process of polymerisation occurs by chaotic condensation of free radicals. For this reason, the structure of each lignin molecule may be unique and hence it is not possible to write it down. However, it is possible to draw a hypothetical partial structure reflecting the different types of phenylpropane building blocks and their different binding pathways.

The lignins of monocotyledonous and dicotyledonous coleopterous and needle plants are structurally different from each other. This difference is based on the unequal structure of phenylpropane building blocks. This is evidenced by different aromatic aldehydes obtained by mild oxidation of lignins from these sources with nitrobenzene in alkaline medium. The lignin of holodiaceous plants yields mainly vanillin and some n-

hydroxybenzaldehyde. Lignin from dicotyledonous plants gives mainly vanillin and lilac aldehyde and a little *p-hydroxybenzaldehyde*; lignin from innodiolaceous plants gives vsetri aldehyde.
There is accumulating evidence in favour of lignin being covalently bound to cell wall matrix polysaccharides, but the nature of this bond has not yet been established.
Lignin is exceptionally difficult to extract from the cell wall. Until relatively recently, all extraction methods involved the use of very aggressive chemicals at elevated temperatures, e.g. 40% HC1, a 3:1 mixture of 36% HC1 and 80% H_2SO_4, 2% H_3RO_4, and then ammonia. SiO. These techniques resulted in significant modification of the extracted lignin. Moreover, the nature of the modification depended on the reagent used for extraction. An additional complication was introduced by self-coideation of lignin molecules, which occurred at elevated temperatures. Such modified lignins were usually named after the extraction reagent used, e.g., hydrochloric acid lignin, cu-proxam lignin.
These difficulties were largely overcome when Browns showed that about 5 per cent of the total wood lignin could be extracted by treating fresh sawdust with 95 per cent ethanol for several weeks at room temperature. This combination of a neutral, relatively inert solvent and low temperature did not cause any appreciable structural changes, and therefore, it was possible to extract about 5 per cent of the total wood lignin by treating fresh sawdust with 95 per cent ethanol over a period of several weeks at room temperature.
The lignin extracted in this way is called Browns lignin or native lignin. The lignin extracted in this way is called Browns lignin or native lignin. The yield of native lignin can be increased by using wood affected by brown rot, such as that caused by the tribe *Pona vaülanrn.* Such fungi secrete a mixture of enzymes that break down cellulose and other wall polysaccharides but do not attack lignin, which becomes more accessible to the extraction solvent. Brown rot These fungi are called brown rot because the affected area of the wood is brown in colour. This distinguishes them from the fungi that cause white rot, which break down all components of wood, including lignin, and leave white rotted areas.
The utilisation of lignocellulosic biomass is based on the process of its conversion into digestible substrates, which in turn serve as raw materials for many industries producing chemicals, fuels and foodstuffs. All components of biomass, cellulose hemicellulose and lignin, must be utilised efficiently. Lignin protects polyglucan from the action of enzymes, so that for successful utilisation of polymeric cellulose it is necessary first of all to remove

hemicellulose, then to destroy the lignin complex": cellulose, and then - and the crystalline structure of cellulose itself. These problems can be solved by applying a variety of chemical, physical and microbiological methods.

The hemicelluloses are easily removed by dissolution in weak acid during the initial processing of biomass. In the production of glucose from wood, hydrolysis with a weak acid solution at high pressure and temperature has been used quite successfully. This method is used in a number of countries. Hydrolysis with concentrated acid is much more efficient, but this process requires large capital investments. Hydrofluoric acid is particularly successful, and anhydrous acid can be regenerated. Lignocellulose can also be milled and then exposed to high doses of radiation. The lignin obtained by these methods varies in its properties. Purified lignins are used in the production of glues, resins, adhesives, printing ink, for dispersing dyes; as .adsorbents, insulators, in oil extraction, and in the application of polymers in asphalt mixtures. After removal of lignin by solvent extraction, crystalline cellulose is obtained, which is used in the paper industry. The profitability of this process is mainly determined by the cost of solvent recovery. An example of a one-stage process under mild conditions is phenol extraction at 100 C, when lignin and hemicelluloses are dissolved in phenol and the pulp released after filtration is used for papermaking. Upon cooling of the phenolic extract, the hemicellulose concentrates mainly in the aqueous phase, separating from the oils and lignin, which can be used to produce new portions of phenol by hydrocracking. Thus, with respect to phenol ^energy requirements, it is a self-sustaining process. During the "vapour explosion" ligiocellulose undergoes transformations that make it more sensitive to enzymatic hydrolysis. At the same time, the yield of reactive lignins released under the action of weak alkali also increases. Liquid ammonia is used for freeze destruction. When moderately hard wood is treated by autogndrolysis on the basis of "steam explosion", waste suitable for further processing by microbes is obtained (coniferous and tropical wood, which is richer in lignin, does not lend itself well to such treatment).

Hydrolysis lignin is an inhomogeneous product of acid processing of wood, which consists of lignin proper (up to 80 %), residues of poly- and monosaccharides, organic acids, resins, waxes, nitrogenous substances, ash elements and mineral acids not washed off during wood hydrolysis.

Scientific and research work to find rational ways of using hydrolysis lignin is carried out by the drolysis production, but the implementation of many proposals due to the complexity and neo-iodicity of the composition of lignin, insufficient logical elaboration of individual stages and apparatus design has not been implemented. For example, technologies for the use of hydrolysis lignin as a component of phenol-formaldehyde resins and rubbers, a sorbent for wastewater treatment have been developed, a technology for obtaining the medicinal drug polyphepan, etc. is being implemented. However, only about 2.7 % of hydrolysed lignin is processed. There are recommendations on the use of hydrolysis lignin as an energy fuel, as the low calorific value of hydrolysis lignin can be an effective measure for its utilisation, but not the reason for using its potential as a natural organic raw material. It is possible to return hydrolysed lignin to the carbon cycle in nature by means of microbiological composting.

There is also a sufficient number of recommendations for the use of hydrolysed lignin mixed with alkaline additives such as manure as a fertiliser. At the same time, there is information about deterioration of soil structure and mechanical composition due to cementing ability of hydrolysed lignin. Obviously, simple mixing of hydrolysed lignin with organic additives is not enough to obtain a complete fertiliser.

One of the directions of processing industrial lignins into opra- but mineral fertilisers is related to the use of microorganisms and their enzymes as delignifying agents, since lignin decomposition is one of the main stages of composting. Due to its complex chemical structure, lignin decomposes extremely slowly. Lignin degradation is mainly carried out by basidial fungi, primarily white rot fungi. Biodegradation of lignin by these fungi is a multistage process involving enzymes of the lignolytic complex.

The problem of environmental pollution by emissions from the pulp and paper and wood-chemical industries is especially acute in areas where hydrolysis enterprises are located, from which a huge amount of hydrolysis lignin is annually landfilled. Such large tonnage wastes occupy considerable land areas and pollute the territories adjacent to the enterprises. Therefore, the main task of biotechnology is not to obtain, but to utilise lignin as a waste product of pulp and paper production.

Chapter 23

UTILISATION OF BIOTECHNOLOGY PRODUCTS FOR ENVIRONMENTAL PROTECTION

Eco-biotechnology solves environmental problems such as waste treatment, water purification, and pollution elimination.

Microorganisms play a huge role in the process of contaminant cycling in ecosystems. However, many of the low-molecular compounds (pesticides, detergents) and high-molecular polymers created by humans have proved to be stable and are not decomposed by microorganisms; moreover, they show mutagenic, carcinogenic and teratogenic effects, so their utilisation requires the development of more advanced treatment technologies.

Typically, micro- haTM (s)complexes and special instrumentation are used for waste management.

Foreign substances (xenobiotics), entering the organism of humans and animals, undergo various biotransformation: oxidation, reduction, hydrolysis and other transformations with the participation of enzyme systems. In water and soil biotransformation of xenobiotics occurs under the influence of enzymes and microorganisms. The study of reactions in soils is complicated by heterogeneity of the medium and adsorption of xenobiotics, microorganisms and enzymes, on soil particles and colloids. Many xenobiotics are quite stable in the biosphere: for example, dichlorodiphenyltrichloroethane (DDT) does not disappear from the soil for 3 years, aldrin and chlordane - 15 years, dieldrin - 25 years, heptachlor - 14 years. Some substances form even more stable and toxic compounds during decomposition.

One of the main aspects of ecobiotechnology is the production of environmentally friendly energy. This is considered to be the energy obtained by converting solar energy into electric energy with the help of solar collectors, as well as biogas and microbial ethanol.

23.1. Biogas production

Biogas is a mixture consisting of 65% methane, 1% carbon dioxide, 1% hydrogen sulphide and minor impurities of nitrogen, oxygen, hydrogen and carbon monoxide. The energy contained in $1m^3$ o

biogas is equivalent to the energy of 0.6 m of natural gas, or 0.74 litres of oil, or 0.66 litres of diesel fuel. Biogas production is based on the process of methane fermentation, or biomethanogenesis, which is the transformation of biomass into energy.

Bnomethanogenesis is a complex microbiological process in which organic matter is decomposed under anaerobic conditions to methane and carbon dioxide. Almost all compounds of natural origin, as well as a significant part of xenobiotics of organic nature, are amenable to microbiological decomposition.

Waste from animal husbandry, agriculture, spoilt products, effluents from starch processing plants, liquid waste from sugar factories, household waste, and municipal wastewater can be used to produce biogas (Fig. 15).

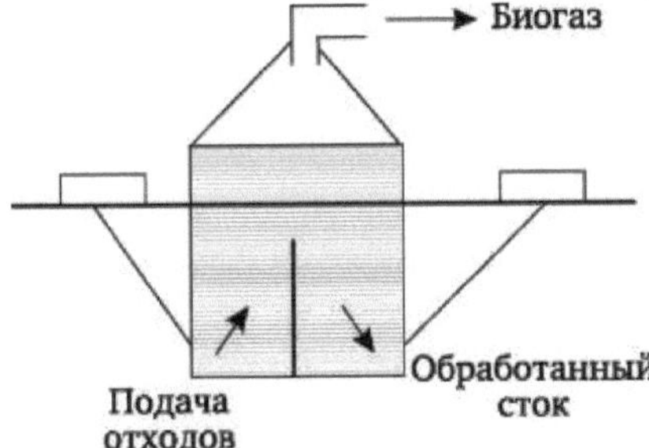

Fig. 15. Schematic diagram of the Agricultural Waste Treatment Reactor device

The process is carried out at a temperature of 30...001 C and pH values from b to 8. Most often secondary wastes are used (i.e. animal waste, municipal water).

Feeding of crop residue manure (substrate) and withdrawal of spent manure (effluent) is carried out in the engineering part of the reactor. The mode of operation can be either batch or semi-continuous. The reactor usually has two (or more) sections to separate the process stages. The biogas is burned to produce carbon dioxide and water, and the natural fertiliser, sapropel, remains in the reactor. It contains nitrogen, phosphorus, potassium salts necessary for plant growth. The use of sapropel is more expedient than manure, as the latter overloads the soil.

In the anaerobic process of biomethanogenesis, there are three successive stages involving more than 190 different microorganisms. In the first stage, complex multi-carbon compounds such as proteins, lipids and polysaccharides undergo enzymatic hydrolysis. The second stage involves the formation of acetate, which can occur in two ways:

. *acetogenic* microorganisms assimilate hydrogen, carbon dioxide and some one-carbon compounds to form acetate;

. go^ceu microorganisms assimilate hydrogen, carbon dioxide and some one-carbon compounds to form acetate.

In the third stage, methai is formed. It can be synthesised through a reduction step of carbon dioxide with molecular hydrogen, or from the methyl, acetate group..Some methane bacteria use formate, carbon dioxide, methanol, methylamine and aromatic compounds as substrates:

$4H_2 + CO_2 \rightarrow CH_4 + 2H_2O$;
$3H_2 + CO \rightarrow CH_4 + H_2O$;
$2H_2O + CO \rightarrow CH_4 + CO_2$;
$4HCOOH \rightarrow CH_4 + 3CO_2$;
$4CH_3OH \rightarrow 3CH_4 + CO_2 + H_2O$;
$CH_3OOH \rightarrow CH_4 + CO_2$.

Depending on the process temperature, methane bacteria are divided into mesophilic and thermophilic bacteria. The optimum temperature for mesophilic bacteria is 30...40 °C, for thermophilic bacteria - 50...60 °C. In general, the thermophilic process of methanogenesis is more intensive than the mesophilic one, and the substrate is decontaminated from pathogenic microflora and helminths.

The microflora for methanogenesis is formed mainly by microflora of the gastro-intestinal tract of animals: *Lactobacillus acidophilus, Eubacterium aerofaciens, Methanobacterium mobile, Methanosarcina* sp., *Methanobrevibacte riumruminantium*.

Methane-forming bacteria convert 90 to 95% of the carbon used into methane and only 5 to 10% into biomass.

Anaerobic bioconversion of organic waste into methane is the most competitive area of bioenergy. It allows to produce from local raw materials biogas as a local energy source. Ecologically clean energy sources do not negatively affect the environment. Modern energy sources - HPPs, TPPs, NPPs - cause serious disturbances in the environment. HPPs cause flooding of territories, change of landscape, death of biocenoses. TPPs pollute the atmosphere, cause land alienation. Nuclear power plants create a threat of radiation pollution. The combustion of oil and gas causes an increase in the concentration of carbon dioxide, smog formation and, moreover, a decrease in the resources of oil and gas.

Biogas is widely used in India, China and Japan. The main advantage of biogas is that it is a renewable source of energy. In nature, about 800 million tonnes of methane are produced annually as a result of bacterial activity and about the same amount is produced by humans.

23.2. Bioethanol production

Since 1975, food ethanol production has remained constant, while fuel ethanol production has increased 10-fold.

Ethanol can be used as a stand-alone fuel™ in a blend of 10 to 26% with petrol (called "gasohol" in the USA), or in a blend of 3% with diesel fuel. These blends can be used without modification of the internal combustion engine design.

Alcohol is used as an energy source in Brazil, USA, EU countries, i.e. in energy deficit zones. In addition, it is widely used in the chemical industry as a solvent, extractant, antifreeze.

Ethanol is produced mainly by digesting sugars from plants.

Alcohol production from sugar cane (Brazil, USA) is the world's largest producer. Sugar cane is carefully crushed, the juice is concentrated and fermented. The second is the production of alcohol from cassava, a starchy plant that can grow in poor soils.

The Brazilian biotechnological solution to the fuel problem was thought to be the most appropriate, but it turned out that the best arable land was planted with sugar cane, requiring about 13,000 metres for one car, compared to 0.8,000 metres per year for one person, i.e. one car takes food from 18 inhabitants.It is known that millions of people in Brazil are malnourished. In addition, effluents from distilleries pollute water bodies and upset the ecological balance.

In addition to sugarcane and cassava, sugarcane, topinambour, pineapple, sugar beet and sorghum are used for alcohol production.

When processing starch-containing raw materials, they need to be pre-digested and treated with enzymes to convert starch and other polysaccharides into microbial digestible sugars.

Molasses - the residue from sugar beet sugar production - and whey from cheese production can also be used to produce alcohol.

Production of alcohol from sugar cane is economically unjustified (Brazil), from corn (USA) - subsidised by the state to keep the price of ethanol lower than petroleum products. In other regions, the cost of bioethanol is even higher because the cost of feedstock is higher. Reducing the cost of ethanol can be achieved by replacing the feedstock or radically changing the fermentation technology.

In the feedstock substitution, instead of cereal grains, biomass of whole plants, both herbaceous and trees, atakkhe municipal solid waste, i.e. rubbish, is used for conversion into ethanol.

Lignocellulose (wood) consists of three polymers: cellulose, hemicellulose and lignin. Through chemical or enzymatic hydrolysis, these polymers are cleaved to monomers, followed by fermentation of sugars to ethanol. Despite incomplete hydrolysis (due to the complexity of the chemical structure of the polymer molecules), the process is economically favourable.

In addition, yeast species capable of digesting alcohol not only hexose (glucose) but also xylose have been found. The use of such yeasts leads to a more complete utilisation of sugars, and consequently, the yield of alcohol is increased and its production cost is reduced.

Hydrolysates of wood and sulphite liquors (a waste product in pulp and paper production) are used to produce technical alcohols in Russia.

23.3. Wastewater treatment

The most important problem of ecological biotechnology is wastewater treatment. The demand for water due to the growth of cities, rapid development of industry, and intensification of agriculture is enormous. Annual water consumption on the globe for all types of water supply is from 3300 to 3500 km^3, with agriculture accounting for 70% of total water consumption. The chemical, pulp and paper, energy, ferrous and coloured metallurgy industries *and the* domestic needs of the population also require significant quantities of water. Most of this water after its use is returned to rivers and lakes in the form of wastewater.

At the present stage, the following directions of rational water resources consumption are highlighted: more complete utilisation and expansion of fresh water resources reproduction; development of new biotechnological processes to prevent pollution of water bodies and minimise the use of fresh water.

Pollution of surface and subsoil water can be divided into several types: mechanical, accompanied by an increase in the content of mechanical impurities and refers mainly to surface types of pollution; chemical, due to the presence in the water of organic and inorganic substances of toxic and non-toxic action; biological, associated with the presence in the water of a variety of pathogenic microorganisms, fungi and algae; radioactive; thermal. The main sources of pollution and contamination of water bodies are insufficiently cleaned wastewater from industrial and municipal enterprises, large livestock complexes, production wastes from the development of ore minerals (water from mines and mines); discharges from water and iron-ore transport; pesticides, etc. Pollutants, getting into natural water bodies, naturally change their composition.

Wastewater from soda, sulphate, nitrogen and tuk plants, lead, zinc, nickel ore processing plants containing acids, alkalis, heavy metal ions change physical properties of water (appearance of unpleasant odours, tastes, etc.). Wastewater from oil refineries, petrochemical plants, organic synthesis enterprises contain various petroleum products, ammonia, aldehydes, resins, phenols and other harmful substances. As a result of oxidative processes, the oxygen content in water decreases and its organic parameters deteriorate.

Oil and petroleum products are the main pollutants of inland water bodies, waters and seas of the World Ocean and create different forms of pollution: oil film floating in the water and heavy fractions on the bottom of water bodies. Water acquires toxic properties and poses a threat to all living things: 12 g of oil makes one tonne of water unusable. A harmful pollutant of industrial water is phenol, contained in the wastewater of many petrochemical plants. Wastewater from the pulp and paper industry has a detrimental effect on the life of the inhabitants of water bodies. Oxidation of wood pulp is accompanied by absorption of a significant amount of oxygen, which leads to the death of eggs, fry and adult fish. Wastewater with high radioactivity (100 curies per 1 litre or more) should be buried in underground drainless basins and special reservoirs.

Synthetic detergents widely used in households, industry and agriculture pollute water bodies to a considerable extent. They paralyse bacteria. Pesticides, getting into water bodies, accumulate in plankton, benthos, fish and through the food chain get into the human body, acting negatively on both individual organs and the body as a whole. Wastewater containing wastes of tannery, pulp and paper industry, sugar and breweries, meat and dairy, canning and confectionery industries cause organic pollution of water bodies. Heated wastewater from thermal power plants causes thermal pollution, which drastically changes the thermal regime and negatively affects the flora and fauna of water bodies. Favourable conditions for mass development of blue-green algae in reservoirs (so-called water blooming) arise.

Application of one or another method of water purification - mechanical, chemical, physico-chemical and biological - in each specific case is determined by the nature and degree of harmfulness of impurities.

The essence of these methods is that mechanical impurities are removed from wastewater by sedimentation and filtration. Depending on the size, coarse particles are captured by grids, sieves, sand traps, manure traps, oil traps, etc. Mechanical treatment makes it possible to extract up to 60-75 % of insoluble impurities from domestic wastewater, and up to 95 % from industrial wastewater, many of which are used in production.

Chemical methods. are related to the fact that various chemical reagents are added to the wastewater, which react with pollutants and precipitate them in the form of insoluble precipitates. Chemical treatment reduces the amount of insoluble impurities up to 95 per cent, apaçTB0PHMMx-fl0 25 per cent.

Physical methods are used for the removal of fine "dissolved inorganic impurities and for the destruction of organic and poorly oxidised substances. They include electrolysis, oxidation, sorption, extraction, ion-exchange chromatography, ultra-sound, high-pressure and hydrogen.

Biological methods are based on the use of regularities of biochemical and physiological, self-purification of rivers and other water bodies. Biofilters, biological ponds and aeration tanks are used for wastewater treatment.

In biofilters, wastewater is passed through layers of coarse-grained material covered with a thin bacterial film, which enables intensive biological oxidation processes to take place. In biological ponds, all organisms inhabiting the water body take part in wastewater treatment.

Aerotanks are huge tanks made of reinforced concrete where treatment is carried out with the help of active sludge made of bacteria and microscopic animals that grow rapidly in these facilities, which is facilitated by organic substances of wastewater and excessive oxygen supplied by the air flow. Bacteria, sticking together in flakes, release enzymes into the environment, destroying the organic contaminants. Sludge with flakes settles down, separating from the treated waterInfusoria, flagellates, amoebae, rotifers and other small animals devouring bacteria that are not stuck together in flakes, thereby rejuvenating the bacterial mass of sludge. Wastewater is first subjected to mechanical and then chemical treatment to remove disease-causing bacteria by chlorination with liquid chlorine or bleach. Ultrasound, ozonation, electrolysis and other methods are also used for disinfection.

The biological method gives significant results in the treatment of municipal and domestic wastewater, as well as waste from oil refining, pulp and paper industry and artificial fibre production.

List of literature

Antibiotics/P.N.Kashkin[et al].L.,1970.

Artamonov, V. I. Amusing physiology of plants / V. I. Artamonov. M., 1991.

Bezborodov, A. M. Biochemical bases of microbiological synthesis /. A.MBezborodov.M., 1984.

Bezborodov, A. M. Enzymes of microorganisms and their application / A. M. Bezborodov // Biotechnology / edited by A. A. Babaev. M., 1984.

Becker, M. E. Biotechnology / M. E. Becker, G. K. Liepinen, E. P. Raipoulis. M., 1990.

Biotechnology / edited by A. A. Babaev. M., 1984

Plant biotechnology: cell culture / A. D. Golovleva [et al].

Biotechnology for agriculture / A. G. Lobanok [et al]. Minsk, 1988.

Biotechnology of agricultural plants / E. V. Levashov [et al]. M., 1987.

v_{ul}0akov, A. Food additives : a reference book / A. Buldakov. SPb., 1996.

Varfolomeev, S.D. Biotechnology of solar energy conversion. Modern state, problems, prospects / S.D. Varfolomeev, E.S. Panz - Biotechnology / ed. by A.A. Babaev. M.,1984.

Varfolomeev, S. D. Biotechnology / S. D. Varfolomeev, S. V. Kalyuzhny. M., 1990.

Viestur, U. E. Biotechnology: Biological agents, technology, apparatus / U. E. Viestur, I. A. Shmita, A. V. Zhilevich. Riga, 1987.

Woodworth, M. Methods of cell culture / M. Woodworth. Л., 1988.

Hemicellulases / M. C. Dudkin [et al.] Riga, 1991.

Glick, B. Molecular technology. Principles and applications / B. Glick, J.Pasternak.M.,2002.

Golubovskaya, E. K. Biological bases of water purification / E. K. GolubovskayaM.1978.

Gracheva, I. M. Technology of microbial protein preparations of amioacids and bioenergy / I. M. Gracheva, L. A. Ivanova, V. M. Cantere. M., 1992.

Gracheva, I. M. Technology of microbial protein preparations, amino acids and fats / I. M. Gracheva, H. M. Gavrilova, L. A. Ivanova. M., 1980.

Gracheva, I. M. Technology of enzyme preparations / I.M. Gracheva, A.Yu.Krivova.Z-e-eizd.M., 2000.

Egorov, N. S. Fundamentals of the doctrine of antibiotics / N. S. Egorov. M., 1986.

Blinov, H. P. Fundamentals of Biotechnology / H. P. Blinov. St. Petersburg, 1995.

Blinov, *N. P.* Chemical microbiology / N. P. Blinov. M., 1989

Immobilised cells and enzymes. Methods / ed. by J. Woodworth-ta-m.-19!(8)-.

Immobilised enzymes / I.V. Berezin [et al]. M., 1987.

Engineering enzymology / IV. Berezin [et al.] M., 1987.

Kkazanskaya, *N. F.* Fermeites and protein preparations in medicine / H F. Kazanskaya, N. I. Larionova, V. P. Torchilin // Biotechnology / ed. by A. A. Babaev M., 1984

Karavaiko, G. I. Biogeotechnology of metals / G. I. Karavaiko // Biotechnology / ed.

by A. A. Babaev. M., 1984.
Kvesitadze, G. I. Introduction to biotechnology / G. I. Kvesitadze, A. M. Me. Bez-borodov. M., 2002.
Kvesitadze, G. I. Fungal and bacterial amylases / G. I. Kvesitadze.
Tbilisi, 1984.
Kefeli, V. I. Biotechnology : a course of lectures / V. I. Kefeli, G. A. Dmitrieva.Puschnio, 1989.
A. A. Application of immobilised enzymes in food industry / A. A. Klesov // Biotechnology / ed. by A. A. Babaev. M., 1984. *Martinek, K.* Immobilised enzymes / K. Martinek // Biotechnology / edited by A.A.Babaev.M., 1984.
Microbiological production of biologically active substances and preparations / edited by N. S. Egorov, V. D. Samuilov. M., 1987.
Molecular biology of the cell : in 3 vols. / B. Albert [et al.] M., 1994. T. 1.
Nechaev, A. P. Food additives / A. P. Nechaev, A. A. Kochetkova, A. N. Zaitsev (2002)-.
Pechurkin, N. S. Population aspects of biotechnology / N. S. Pechurkin, A. V. Brilkov, T. V. Marchenkova. Novosibirsk, 1990.
Piruzyan L. A. A. Saprotrophic microflora as a producer of biologically active substances for the purposes of microbial saprotrophy and pharmacotherapy / L. A. Piruzyan, E. M. Mihailovskii. Mikhailovskii // Izv. RAS. Ser. biol. 1992. № 6. (c)- "(6)- -
Production of protein substances / V. A. Bykov[et al]. M.,1987
Industrial microbiology / ed. by N. S. Egorov M., 1989.
Renneberg, R. From bakery to biofactory / R. Renneberg, I. Renneberg. M., 1991.
Rychkov, R. *S.* Biotechnology: prospects of development / R. S. Rychkov, V. G. Po-pov // Biotechnology / edited by A. A. Babaev.M., 1984.
Savitskaya, E. I. Biology of our days / E. I. Savitskaya. M., 1987. Vol. 2.
Samartsev, M. A. Application of immobilised enzymes in industrial processes/M. A. Samartsev, N.V. Belyakov, A.I. Kestner. M., 1984.
Sasson, A. Biotechnology: achievements and hopes / A. Sasson. M., 1987.
Scopes, R. Methods of protein purification / R. Scopes M., 1985.
Skryabin, G. K. Immobilised cells of microorganisms / G. K. Skryabin, K. ALSoscheiko // Biotechnology / ed. by A. A. Babaev. M., 1984
Technology of fat processing / edited by N. S. Arutyunyan. M., 1985.
Technology of products from hydrobionts / ed. by T. M. Safronova,
B. I.Shsndsryuk. M., 2001.
Technologiya alcohola / ed. by V.L. Yaroveshu M., 2002.
Triven, M. Immobilised enzymes / M. Triven. M., 1983.
Tyutyunnikov, B. N Chemistry of fats / B. N. Tyutyunnikov. M.1966.
Enzyme preparations in animal feeding / L. G. Boyarskii [et al.].
(M)--1.(985)-..
Higgins, I. Biotechnology. Principles and Applications / I. Higgins, D. Best, J.

Jones, M., 1988.
Khotnovich, A. V. Methods of cultivation of nitrogen-fixing bacteria, methods of obtaining and use of preparations on their basis (methodological recommendations)/A. V. Kh*otnovich*. Л., 1991
Schlegel, G. General microbiology / G. Schlegel. M., 1987.

Recommended websites for self-study

www.antibiotic.ru www.biengi.ac.ru www.bio.lseptember.ru www.bioinforum.ru www.biolinks.net.ru www.bioscience.ru www.chem.ac.ru www.chemistry.narod.ru www.chshb.ru www.edu.ru www.informnauka.ru

www.markovsky.virtualate.neywww.medline.ru www.medlinks.ru www.mednovosti.ru www.membrana.ru www.molbio.ru www.nature.ru www.rusbiotech.ru www.scientific.ru

Printed by Books on Demand GmbH, Norderstedt / Germany